Student Soluti

John Garlow

John Tobey Jeffrey Slater

Basic College Mathematics

Fifth Edition

PEARSON
Prentice Hall

Upper Saddle River, New Jersey 07458

Editor-in-Chief, Developmental Math: Chris Hoag
Senior Acquisitions Editor: Paul Murphy
Supplements Editor: Christina Simoneau
VP of Production and Manufacturing: David W. Riccardi
Executive Managing Editor: Kathleen Schiaparelli
Managing Editor: Becca Richter
Production Editor: Zachary Hubert
Supplement Cover Manager: Daniel Sandin
Manufacturing Buyer: Ilene Kahn

Pearson Prentice Hall
Pearson Education, Inc.
Upper Saddle River, NJ 07458

Printed in the United States of America

10 9 8 7 6 5 4 3 2 1

ISBN 0-13-149061-3

Pearson Education Ltd., *London*
Pearson Education Australia Pty. Ltd., *Sydney*
Pearson Education Singapore, Pte. Ltd.
Pearson Education North Asia Ltd., *Hong Kong*
Pearson Education Canada, Inc., *Toronto*
Pearson Educación de Mexico, S.A. de C.V.
Pearson Education—Japan, *Tokyo*
Pearson Education Malaysia, Pte. Ltd.

Table of Contents

Chapter 1

1.1 Exercises

1. $6731 = 6000 + 700 + 30 + 1$

3. $108,276 = 100,000 + 8000 + 200 + 70 + 6$

5. $23,761,345$
$= 20,000,000 + 3,000,000 + 700,000$
$+ 60,000 + 1000 + 300 + 40 + 5$

7. $103,260,768$
$= 100,000,000 + 3,000,000 + 200,000$
$+ 60,000 + 700 + 60 + 8$

9. $600 + 70 + 1 = 671$

11. $9000 + 800 + 60 + 3 = 9863$

13. $40,000 + 800 + 80 + 5 = 40,885$

15. $700,000 + 6,000 + 200 = 706,200$

17. a. 7 b. 50,000

19. a. 2 b. 200,000

21. 53 = Fifty-three

23. 9304 = nine thousand, three hundred four

25. 36,118 = thirty-six thousand, one hundred eighteen

27. 105,261 = one hundred five thousand, two hundred sixty-one

29. 14,203,326 = fourteen million,
two hundred three thousand, three hundred twenty-six

31. 4,302,156,200 = four billion, three hundred two million, one hundred fifty-six thousand, two hundred

33. 1561

35. 27,382

37. 100,079,826

39. one thousand, nine hundred sixty-five

41. 9 million or 9,000,000

43. 33 million or 33,000,000

45. 930,000

47. 52,566,000

49. a. 5
b. 2

51. a. 8
b. 9

53. 613,001,033,208,003

55. Three quintillion, six hundred eighty-two quadrillion, nine hundred sixty-eight trillion, nine billion, nine hundred thirty-one million, nine hundred sixty thousand, seven hundred forty-seven

57. 2E20 or 200,000,000,000,000,000,000

1.2 Exercises

1. Answers may vary. Samples are below.
a. You can change the order of the addends without changing the sum.
b. You can group the addends in any way without changing the sum.

3.

+	3	5	4	8	0	6	7	2	9	1
2	5	7	6	10	2	8	9	4	11	3
7	10	12	11	15	7	13	14	9	16	8
5	8	10	9	13	5	11	12	7	14	6
3	6	8	7	11	3	9	10	5	12	4
0	3	5	4	8	0	6	7	2	9	1
4	7	9	8	12	4	10	11	6	13	5
1	4	6	5	9	1	7	8	3	10	2
8	11	13	12	16	8	14	15	10	17	9
6	9	11	10	14	6	12	13	8	15	7
9	12	14	13	17	9	15	16	11	18	10

5. 4
2
8
+ 9
23

7. 2
6
7
8
+ 3
26

9. 18
36
+ 3
57

11. 63
24
+ 12
99

13. 2847
1634
+ 98
4579

15. 5631
2344
+ 2019
9994

17. 8235
+ 5626
13,861

19. 62,504
+ 54,736
117,240

21. 36
41
25
6
+ 13
121

23. 207
15
3
57
+ 861
1143

25. 85
256
55
+ 9734
10,130

27. 1,362,214
7,002,316
+ 3,214,896
11,579,426

29.
$$\begin{array}{r} 837,241,000 \\ +\ 298,039,240 \\ \hline 1,135,280,240 \end{array}$$

31.
$$\begin{array}{r} 516,208 \\ 24,317 \\ +\ 1,763,295 \\ \hline 2,303,820 \end{array}$$

33.
$$\begin{array}{r} 75 \\ 132 \\ 25 \\ +\ 51 \\ \hline 283 \end{array}$$

35.
$$\begin{array}{r} 15,216 \\ 485 \\ +\ 5208 \\ \hline 20,909 \end{array}$$

37.
$$\begin{array}{r} 345 \\ 288 \\ +\ 74 \\ \hline 707 \end{array}$$

The cost was $707.

39.
$$\begin{array}{r} 4550 \\ 9200 \\ +\ 6875 \\ \hline 20,625 \end{array}$$

She made $20,625.

41.
$$\begin{array}{r} 124 \\ 105 \\ 147 \\ +\ 92 \\ \hline 468 \end{array}$$

468 ft of fencing are needed.

43.
$$\begin{array}{r} 64,000,000 \\ 31,800,000 \\ +\ 25,300,000 \\ \hline 121,100,000 \end{array}$$

The total area is 121,100,000 square miles.

45.
$$\begin{array}{r} 3,072,149 \\ 843,746 \\ +\ 179,727 \\ \hline 4,095,622 \end{array}$$

4,095,622 people.

47. a.
$$\begin{array}{r} 414 \\ 364 \\ 159 \\ +\ 196 \\ \hline 1134 \end{array}$$
1134 students

b.
$$\begin{array}{r} 1134 \\ 27 \\ 68 \\ 102 \\ +\ 61 \\ \hline 1392 \end{array}$$
1392 students

49.
$$\begin{array}{r} 87 \\ 17 \\ +\ 98 \\ \hline 202 \end{array}$$
202 miles

51.
$$\begin{array}{r} 568 \\ 682 \\ +\ 703 \\ \hline 1953 \end{array}$$

Then,
$$\begin{array}{r} 2387 \\ -\ 1953 \\ \hline 434 \end{array}$$
434 feet

53. a. Out-of state
$$\begin{array}{r} 5276 \\ 2437 \\ +\ 1840 \\ \hline \$9553 \end{array}$$

b. In-state
$$\begin{array}{r} 3640 \\ 1926 \\ +\ 1753 \\ \hline \$7319 \end{array}$$

c. Foreign
$$\begin{array}{r} 8352 \\ 2855 \\ +\ 1840 \\ \hline \$13,047 \end{array}$$

55. $89+166+23+45+72+190+203+77+18$
$+93+46+73+66=1161$

57. Answers may vary. A sample is:
You could not group the addends in groups that sum to 10s to make column addition easier.

Cumulative Review

59. 121,000,374 = One hundred twenty-one million, three hundred seventy-four.

61. Nine million, fifty-one thousand, seven hundred nineteen = 9,051,719

1.3 Exercises

3. In subtraction, the minuend minus the subtrahend equals the difference. To check, we add the subtrahend and the difference to see if we get the minuend. If we do, the answer is correct.

3. We know that $1683 + 1592 = 32?5$. Therefore, If we add 8 tens and 9 tens, we get 17 tens which is 1 hundred and 7 tens. Thus the ? should be replaced by 7.

5. $\begin{array}{r} 8 \\ -\ 3 \\ \hline 5 \end{array}$

7. $\begin{array}{r} 15 \\ -\ 9 \\ \hline 6 \end{array}$

9. $\begin{array}{r} 16 \\ -\ 0 \\ \hline 16 \end{array}$

11. $\begin{array}{r} 18 \\ -\ 9 \\ \hline 9 \end{array}$

13. $\begin{array}{r} 11 \\ -\ 4 \\ \hline 7 \end{array}$

15. $\begin{array}{r} 13 \\ -\ 7 \\ \hline 6 \end{array}$

17. $\begin{array}{r} 11 \\ -\ 8 \\ \hline 3 \end{array}$

19. $\begin{array}{r} 15 \\ -\ 6 \\ \hline 9 \end{array}$

21. $\begin{array}{r} 47 \\ -\ 26 \\ \hline 21 \end{array}$

Check: $\begin{array}{r} 26 \\ +\ 21 \\ \hline 47 \end{array}$

23. $\begin{array}{r} 85 \\ -\ 73 \\ \hline 12 \end{array}$

Check: $\begin{array}{r} 73 \\ +\ 12 \\ \hline 85 \end{array}$

25. $\begin{array}{r} 189 \\ -\ 65 \\ \hline 124 \end{array}$

Check: $\begin{array}{r} 65 \\ +\ 124 \\ \hline 189 \end{array}$

27. $\begin{array}{r} 869 \\ -\ 548 \\ \hline 321 \end{array}$ $\begin{array}{r} 548 \\ +\ 321 \\ \hline 869 \end{array}$

29. $\begin{array}{r} 2893 \\ -\ \ 572 \\ \hline 2321 \end{array}$

Check: $\begin{array}{r} 2321 \\ +\ \ 572 \\ \hline 2893 \end{array}$

31. $\begin{array}{r} 155,835 \\ -\ \ 12,600 \\ \hline 143,235 \end{array}$

Check: $\begin{array}{r} 12,600 \\ +\ 143,235 \\ \hline 155,835 \end{array}$

33. $\begin{array}{r} 986,302 \\ -\ 433,201 \\ \hline 553,101 \end{array}$

Check: $\begin{array}{r} 553,101 \\ +\ 433,201 \\ \hline 986,302 \end{array}$

35. $\begin{array}{r} 129 \\ -\ \ 19 \\ \hline 110 \end{array}$ $\begin{array}{r} 19 \\ +\ 110 \\ \hline 129 \end{array}$

Correct

37. $\begin{array}{r} 8596 \\ +\ 3215 \\ \hline 5781 \end{array}$ $\begin{array}{r} 3215 \\ -\ 5781 \\ \hline 8996 \end{array}$

Incorrect

Correct answer: 5381

39. $\begin{array}{r} 6030 \\ -\ 5020 \\ \hline 1020 \end{array}$ $\begin{array}{r} 5020 \\ +\ 1020 \\ \hline 6040 \end{array}$

Incorrect

Correct answer: 1010

41. $\begin{array}{r} 47,869 \\ +\ 33,846 \\ \hline 13,023 \end{array}$ $\begin{array}{r} 33,846 \\ +\ 13,023 \\ \hline 46,869 \end{array}$

Incorrect

Correct answer: 14,023

43. $\begin{array}{r} 98 \\ -\ 52 \\ \hline 46 \end{array}$

45. $\begin{array}{r} 174 \\ -\ \ 82 \\ \hline 92 \end{array}$

47. $\begin{array}{r} 647 \\ -\ 263 \\ \hline 384 \end{array}$

49. $\begin{array}{r} 955 \\ -\ 237 \\ \hline 718 \end{array}$

51. $\begin{array}{r} 30,000 \\ -\ 19,370 \\ \hline 10,630 \end{array}$

53. $\begin{array}{r} 152,000 \\ -\ 117,908 \\ \hline 34,092 \end{array}$

55. $\begin{array}{r} 45,312 \\ -\ 37,865 \\ \hline 7447 \end{array}$

57. $\begin{array}{r} 2,378,862 \\ -\ 1,469,932 \\ \hline 908,930 \end{array}$

59. $x + 14 = 19$

$5 + 14 = 19$

$x = 5$

61. $57 = x + 28$

$57 = 29 + 28$

$x = 29$

63. $100 + x = 127$

$100 + 27 = 127$

$x = 27$

65.
$$\begin{array}{r} 960 \\ -\ 778 \\ \hline 182 \end{array}$$
votes

67.
$$\begin{array}{r} 10,066,523 \\ -\ 3,841,268 \\ \hline 6,225,255 \end{array}$$
less people

69. Total earned = \$1280

Total paid out = \$318 + \$200 = \$518

1280 − 518 = \$762 put in checking

71.
$$\begin{array}{r} 4,830,784 \\ -\ 3,413,864 \\ \hline 1,416,920 \end{array}$$
people

73. Illinois = 10,081,158

Indiana + Minnesota = 4,662,498 + 3,413,864

= 8,076,362

Difference = 10,081,158 − 8,076,362

= 2,004,796 people

75.
$$\begin{array}{r} 11,430,602 \\ -\ 11,110,285 \\ \hline 320,317 \end{array}$$
people

77.
$$\begin{array}{r} 413,471 \\ -\ 320,317 \\ \hline 93,154 \end{array}$$
people

79.
$$\begin{array}{r} 125 \\ -\ 96 \\ \hline 29 \end{array}$$
homes

81.
$$\begin{array}{r} 219 \\ -\ 139 \\ \hline 80 \end{array}$$
homes

83. Between 2002 and 2003

85. Willow Creek and Irving

87. It is true if a and b represent the same number, for example, if $a = 10$ and $b = 10$.

89.
$$\begin{array}{r} 276 \\ -\ 216 \\ \hline 60 \end{array}$$

Now, $\frac{60}{12} = 5$ and

$$\begin{array}{r} 110 \\ \times\ \ 5 \\ \hline \$\ 550 \end{array}$$

Cumulative Review

91. Eight million, four hundred sixty-six thousand, eighty-four = 8,466,084

93.
$$\begin{array}{r} 25 \\ 75 \\ 80 \\ 20 \\ +\ 18 \\ \hline 218 \end{array}$$

1.4 Exercises

1. Answers may vary. Samples are below
   a. You can change the order of the factors without changing the product.
   b. You can group the factors in any way without changing the product.

3.

×	6	2	3	8	0	5	7	9	12	4
5	30	10	15	40	0	25	35	45	60	20
7	42	14	21	56	0	35	49	63	82	28
1	6	2	3	8	0	5	7	9	12	4
0	0	0	0	0	0	0	0	0	0	0
6	36	12	18	48	0	30	42	54	72	24
2	12	4	6	16	0	10	14	18	24	8
3	18	6	9	24	0	15	21	27	36	12
8	48	16	24	64	0	40	56	72	96	32
4	24	8	12	32	0	20	28	36	48	16
9	54	18	27	72	0	45	63	81	108	36

5. $\begin{array}{r} 32 \\ \times\ 3 \\ \hline 96 \end{array}$

7. $\begin{array}{r} 14 \\ \times\ 5 \\ \hline 70 \end{array}$

9. $\begin{array}{r} 87 \\ \times\ 6 \\ \hline 522 \end{array}$

11. $\begin{array}{r} 231 \\ \times\ 3 \\ \hline 693 \end{array}$

13. $\begin{array}{r} 429 \\ \times\ 8 \\ \hline 3432 \end{array}$

15. $\begin{array}{r} 6102 \\ \times\ 3 \\ \hline 18,306 \end{array}$

17. $\begin{array}{r} 12,203 \\ \times\ 3 \\ \hline 36,609 \end{array}$

19. $\begin{array}{r} 5218 \\ \times\ 6 \\ \hline 31,308 \end{array}$

21. $\begin{array}{r} 12,526 \\ \times\ 8 \\ \hline 100,208 \end{array}$

23. $\begin{array}{r} 235,702 \\ \times\ 4 \\ \hline 942,808 \end{array}$

25. $\begin{array}{r} 156 \\ \times\ 10 \\ \hline 1560 \end{array}$

27. $\begin{array}{r} 27,158 \\ \times\ 100 \\ \hline 2,715,800 \end{array}$

29. $\begin{array}{r} 482 \\ \times\ 1000 \\ \hline 482,000 \end{array}$

31. $\begin{array}{r} 37,256 \\ \times\ 10,000 \\ \hline 372,560,000 \end{array}$

33. $\begin{array}{r} 423 \\ \times\ 20 \\ \hline 8460 \end{array}$

35. $\begin{array}{r} 2120 \\ \times\ 30 \\ \hline 63,600 \end{array}$

37. $\begin{array}{r} 14,000 \\ \times\ 4000 \\ \hline 56,000,000 \end{array}$

39.
$$\begin{array}{r} 514 \\ \times\ \ 12 \\ \hline 1028 \\ 514 \\ \hline 6168 \end{array}$$

41.
$$\begin{array}{r} 146 \\ \times\ \ 54 \\ \hline 583 \\ 730 \\ \hline 7884 \end{array}$$

43.
$$\begin{array}{r} 89 \\ \times\ 64 \\ \hline 356 \\ 534 \\ \hline 5696 \end{array}$$

45.
$$\begin{array}{r} 607 \\ \times\ \ \ 25 \\ \hline 3\ 035 \\ 12\ 14 \\ \hline 15{,}175 \end{array}$$

47.
$$\begin{array}{r} 659 \\ \times\ \ 67 \\ \hline 4613 \\ 3954 \\ \hline 44{,}153 \end{array}$$

49.
$$\begin{array}{r} 912 \\ \times\ \ 76 \\ \hline 5472 \\ 6384 \\ \hline 69{,}312 \end{array}$$

51.
$$\begin{array}{r} 5123 \\ \times\ \ \ \ 29 \\ \hline 46\ 107 \\ 102\ 46 \\ \hline 148{,}567 \end{array}$$

53.
$$\begin{array}{r} 9053 \\ \times\ \ \ \ 91 \\ \hline 9\ 053 \\ 814\ 77 \\ \hline 823{,}823 \end{array}$$

55.
$$\begin{array}{r} 5536 \\ \times\ \ \ 224 \\ \hline 22\ 144 \\ 110\ 72 \\ 1\ 107\ 2 \\ \hline 1{,}240{,}064 \end{array}$$

57.
$$\begin{array}{r} 678 \\ \times\ \ 132 \\ \hline 1\ 356 \\ 20\ 34 \\ 67\ 8 \\ \hline 89{,}496 \end{array}$$

59.
$$\begin{array}{r} 2076 \\ \times\ \ \ 105 \\ \hline 10\ 380 \\ 00\ 00 \\ 207\ 6 \\ \hline 217{,}980 \end{array}$$

61.
$$\begin{array}{r} 3561 \\ \times\ \ \ 403 \\ \hline 10\ 683 \\ 00\ 00 \\ 1\ 424\ 4 \\ \hline 1{,}435{,}083 \end{array}$$

63. $\begin{array}{r} 12{,}000 \\ \times \quad 60 \\ \hline 720{,}000 \end{array}$

65. $\begin{array}{r} 250 \\ \times \quad 40 \\ \hline 10{,}000 \end{array}$

67. $\begin{array}{r} 302 \\ \times \quad 300 \\ \hline 90{,}600 \end{array}$

69. $7 \cdot 2 \cdot 5 = 7 \cdot 10 = 70$

71. $11 \cdot 7 \cdot 4 = 77 \cdot 4 = 308$

73. $\begin{array}{r} 576 \\ \times \quad 32 \\ \hline 1\,152 \\ 17\,28 \\ \hline 18{,}432 \end{array}$

75. $\begin{aligned} 5 \cdot 8 \cdot 4 \cdot 16 &= 40 \cdot 4 \cdot 10 \\ &= 160 \cdot 10 \\ &= 1600 \end{aligned}$

77. $x = 0$

79. $\begin{array}{r} 38 \\ \times\ 20 \\ \hline 760 \end{array}$ square feet

81. $12 \cdot 14 = 168$

$9 \cdot 3 = 27$

$168 + 27 = 195$ square feet

83. $\begin{array}{r} 240 \\ \times \quad 5 \\ \hline \$1200 \end{array}$

85. $\begin{array}{r} 266 \\ \times \quad 12 \\ \hline 532 \\ 266 \\ \hline \$3192 \end{array}$

87. $\begin{array}{r} 34 \\ \times\ 18 \\ \hline 272 \\ 34 \\ \hline 612 \end{array}$ miles

89. $\begin{array}{r} 485 \\ \times \quad 14 \\ \hline 1940 \\ 485 \\ \hline \$6790 \end{array}$

91. $\begin{array}{r} 6{,}890{,}000 \\ \times \qquad 1070 \\ \hline 482\,300\,000 \\ 000\,000\,0 \\ 6\,890\,000 \\ \hline \$7{,}372{,}300{,}000 \end{array}$

93. $\begin{array}{r} 18 \\ \times\ 2 \\ \hline 36 \end{array}$ $\begin{array}{r} 26 \\ \times 0 \\ \hline 0 \end{array}$ $\begin{array}{r} 54 \\ \times 3 \\ \hline 162 \end{array}$ $\begin{array}{r} 36 \\ 0 \\ +\ 162 \\ \hline 198 \end{array}$ black paws

95. $\begin{array}{r} 18 \\ \times\ 2 \\ \hline 36 \end{array}$ $\begin{array}{r} 26 \\ \times 1 \\ \hline 26 \end{array}$ $\begin{array}{r} 54 \\ \times 0 \\ \hline 0 \end{array}$ $\begin{array}{r} 36 \\ 26 \\ +\ 0 \\ \hline 62 \end{array}$ black ears

97. $5(x) = 40$

$5(8) = 40$

$x = 8$

99. $72 = 8(x)$
$72 = 8(9)$
$x = 9$

101. No, it would not always be true. In our number system $62 = 60 + 2$. But in roman numerals, IV $\neq$ I + V. The digit system in roman numerals involves subtraction. Thus (XII) $\times$ (IV) $\neq$ (XII $\times$ I) + (XII $\times$ V).

Cumulative Review

103.
$$\begin{array}{r} 34,084 \\ -\ 27,328 \\ \hline 6,756 \end{array}$$

105. $156 - (12 + 2 + 3) = 156 - 17 = \$139$

107.
$$\begin{array}{r} 34,005 \\ -\ 32,176 \\ \hline 1,829 \end{array}$$ people

1.5 Exercises

1. a. When you divide a nonzero number by itself, the result is 1.
b. When you divide a number by 1, the result is the number itself.
c. When you divide zero by a nonzero number, the result is zero.
d. You cannot divide a number by zero. Division by 0 is undefined.

3. $6\overline{)42}$ = 7

5. $8\overline{)24}$ = 3

7. $8\overline{)40}$ = 5

9. $9\overline{)36}$ = 4

11. $7\overline{)21}$ = 3

13. $8\overline{)56}$ = 7

15. $7\overline{)63}$ = 9

17. $8\overline{)72}$ = 9

19. $9\overline{)63}$ = 7

21. $6\overline{)24}$ = 4

23. $10\overline{)0}$ = 0

25. $9 \div 0$
undefined

27. $8\overline{)0}$ = 0

29. $6 \div 6 = 1$

31. $6\overline{)29}$ = 4 R5
$$\begin{array}{r} 29 \\ \underline{24} \\ 5 \end{array}$$

Check:
$$\begin{array}{r} 6 \\ \times 4 \\ \hline 24 \\ +5 \\ \hline 29 \end{array}$$

33. $\begin{array}{r} 9 \text{ R4} \\ 8\overline{)76} \\ \underline{72} \\ 4 \end{array}$

Check: $\begin{array}{r} 8 \\ \underline{\times 9} \\ 72 \\ \underline{+4} \\ 76 \end{array}$

35. $\begin{array}{r} 25 \text{ R3} \\ 5\overline{)128} \\ \underline{125} \\ 3 \end{array}$

Check: $\begin{array}{r} 25 \\ \underline{\times\ 5} \\ 125 \\ \underline{+\ 3} \\ 128 \end{array}$

37. $\begin{array}{r} 21 \text{ R7} \\ 9\overline{)196} \\ \underline{18} \\ 16 \\ \underline{9} \\ 7 \end{array}$

Check: $\begin{array}{r} 21 \\ \underline{\times\ 9} \\ 189 \\ \underline{+\ 7} \\ 196 \end{array}$

39. $\begin{array}{r} 32 \\ 9\overline{)288} \\ \underline{27} \\ 18 \\ \underline{18} \\ 0 \end{array}$

Check: $\begin{array}{r} 32 \\ \underline{\times 9} \\ 288 \end{array}$

41. $\begin{array}{r} 37 \\ 5\overline{)185} \\ \underline{15} \\ 35 \\ \underline{35} \\ 0 \end{array}$

Check: $\begin{array}{r} 37 \\ \underline{\times 5} \\ 185 \end{array}$

43. $\begin{array}{r} 322 \text{ R1} \\ 4\overline{)1289} \\ \underline{12} \\ 8 \\ \underline{8} \\ 9 \\ \underline{8} \\ 1 \end{array}$

45. $\begin{array}{r} 127 \text{ R1} \\ 6\overline{)763} \\ \underline{6} \\ 16 \\ \underline{12} \\ 43 \\ \underline{42} \\ 1 \end{array}$

47. $\begin{array}{r} 753 \\ 8\overline{)6024} \\ \underline{56} \\ 42 \\ \underline{40} \\ 24 \\ \underline{24} \\ 0 \end{array}$

49.
$$\begin{array}{r} 1122 \text{ R1} \\ 3\overline{)3367} \\ \underline{3} \\ 3 \\ \underline{3} \\ 6 \\ \underline{6} \\ 7 \\ \underline{6} \\ 1 \end{array}$$

51.
$$\begin{array}{r} 2\,056 \text{ R2} \\ 8\overline{)16,450} \\ \underline{16} \\ 45 \\ \underline{40} \\ 50 \\ \underline{48} \\ 2 \end{array}$$

53.
$$\begin{array}{r} 2\,562 \text{ R3} \\ 5\overline{)12,813} \\ \underline{10} \\ 28 \\ \underline{25} \\ 31 \\ \underline{30} \\ 13 \\ \underline{10} \\ 3 \end{array}$$

55.
$$\begin{array}{r} 30 \text{ R5} \\ 6\overline{)185} \\ \underline{18} \\ 5 \\ \underline{0} \\ 5 \end{array}$$

57.
$$\begin{array}{r} 5 \text{ R7} \\ 52\overline{)267} \\ \underline{260} \\ 7 \end{array}$$

59.
$$\begin{array}{r} 7 \\ 61\overline{)427} \\ \underline{427} \\ 0 \end{array}$$

61.
$$\begin{array}{r} 418 \text{ R8} \\ 12\overline{)5024} \\ \underline{48} \\ 24 \\ \underline{12} \\ 104 \\ \underline{96} \\ 8 \end{array}$$

63.
$$\begin{array}{r} 48 \text{ R12} \\ 30\overline{)1452} \\ \underline{120} \\ 252 \\ \underline{240} \\ 12 \end{array}$$

65.
$$\begin{array}{r} 327 \\ 8\overline{)2616} \\ \underline{24} \\ 21 \\ \underline{16} \\ 56 \\ \underline{56} \\ 0 \end{array}$$

67. $\begin{array}{r} 210 \text{ R8} \\ 36\overline{)7568} \\ \underline{72} \\ 36 \\ \underline{36} \\ 8 \\ \underline{0} \\ 8 \end{array}$

69. $\begin{array}{r} 14 \text{ R2} \\ 182\overline{)2550} \\ \underline{182} \\ 730 \\ \underline{728} \\ 2 \end{array}$

71. $\begin{array}{r} 4 \text{ R4} \\ 174\overline{)700} \\ \underline{696} \\ 4 \end{array}$

73. $\begin{array}{r} 16 \\ 132\overline{)2112} \\ \underline{132} \\ 792 \\ \underline{792} \\ 0 \end{array}$

75. $\begin{array}{r} 37 \\ 14\overline{)518} \\ \underline{42} \\ 98 \\ \underline{98} \\ 0 \end{array}$ Thus, $518 \div 14 = 37$

$x = 37$

77. $\begin{array}{r} 61{,}693 \\ 7\overline{)431{,}851} \\ \underline{42} \\ 11 \\ \underline{7} \\ 48 \\ \underline{42} \\ 65 \end{array}$

61,693 runs

79. $\begin{array}{r} 154 \\ 15\overline{)2310} \\ \underline{15} \\ 81 \\ \underline{75} \\ 60 \\ \underline{60} \\ 0 \end{array}$

154 guests

81. $\begin{array}{r} 21{,}053 \\ 7\overline{)147{,}371} \\ \underline{14} \\ 7 \\ \underline{7} \\ 37 \\ \underline{35} \\ 21 \\ \underline{21} \\ 0 \end{array}$

\$21,053 per carriage

83.
$$\begin{array}{r} 12{,}140 \\ 15\overline{)182{,}100} \\ \underline{15}\phantom{0{,}100} \\ 32\phantom{{,}100} \\ \underline{30}\phantom{{,}100} \\ 21 \\ \underline{15} \\ 60 \\ \underline{60} \\ 0 \end{array}$$

\$12,140

85. The smallest number is 330 and $360 \div 2 = 165$ sandwiches.

87. a. $2 \times 12 = 24$

$$\begin{array}{r} 1742 \\ \times \quad 24 \\ \hline 6968 \\ 3484 \\ \hline 41{,}808 \end{array}$$ kilometers

b.
$$\begin{array}{r} 50{,}000 \\ -\ 41{,}808 \\ \hline 8192 \end{array}$$ kilometers

89. a and b must represent the same number. For example, if $a = 12$, then $b = 12$.

Cumulative Review Problems

91.
$$\begin{array}{r} 108 \\ \times \ 50 \\ \hline 5400 \end{array}$$

93.
$$\begin{array}{r} 316{,}214 \\ +\ \ 89{,}981 \\ \hline 406{,}195 \end{array}$$

How Am I Doing? Sections 1.1-1.5

1. 78,310,436 = seventy-eight million, three hundred ten thousand, four hundred thirty-six

2. $38{,}247 = 30{,}000 + 8000 + 200 + 40 + 7$

3. $5{,}064{,}122$

4. $2{,}747{,}000$

5. $2{,}583{,}000$

6.
$$\begin{array}{r} 13 \\ 31 \\ 88 \\ 43 \\ +\ 69 \\ \hline 244 \end{array}$$

7.
$$\begin{array}{r} 28{,}318 \\ 5{,}039 \\ +\ 17{,}213 \\ \hline 50{,}570 \end{array}$$

8.
$$\begin{array}{r} 7{,}148 \\ 500 \\ 19 \\ +\ 7{,}062 \\ \hline 14{,}729 \end{array}$$

9.
$$\begin{array}{r} 6439 \\ -\ 2689 \\ \hline 3750 \end{array}$$

10.
$$\begin{array}{r} 100{,}450 \\ -\ 24{,}139 \\ \hline 76{,}311 \end{array}$$

11.
$$\begin{array}{r} 45{,}861{,}413 \\ -\ 43{,}879{,}761 \\ \hline 1{,}981{,}652 \end{array}$$

12. $9 \times 6 \times 1 \times 2 = 54 \times 1 \times 2 = 54 \times 2 = 108$

13. $3200 \times 40 \times 10 = 128,000 \times 10 = 1,280,000$

14.
$$\begin{array}{r} 2658 \\ \times \quad 7 \\ \hline 18,606 \end{array}$$

15.
$$\begin{array}{r} 91 \\ \times \ 74 \\ \hline 364 \\ 637\ \, \\ \hline 6734 \end{array}$$

16.
$$\begin{array}{r} 365 \\ \times \ 908 \\ \hline 2\ 920 \\ 328\ 50 \\ \hline 331,420 \end{array}$$

17.
$$\begin{array}{r} 10,605 \\ 8\overline{)84,840} \\ \underline{8} \\ 4\ 8 \\ \underline{4\ 8} \\ 40 \\ \underline{40} \\ 0 \end{array}$$

18.
$$\begin{array}{r} 7,376 \text{ R1} \\ 7\overline{)51,633} \\ \underline{49} \\ 2\ 6 \\ \underline{2\ 1} \\ 53 \\ \underline{49} \\ 43 \\ \underline{42} \\ 1 \end{array}$$

19.
$$\begin{array}{r} 26 \\ 76\overline{)1976} \\ \underline{152} \\ 456 \\ \underline{456} \\ 0 \end{array}$$

20.
$$\begin{array}{r} 139 \\ 42\overline{)5838} \\ \underline{42} \\ 163 \\ \underline{126} \\ 378 \\ \underline{378} \\ 0 \end{array}$$

1.6 Exercises

1. 5^3 means $5 \times 5 \times 5 = 125$

3. base

5. To insure consistency we
1. Perform operations inside parentheses.
2. Simplify any expressions with exponents.
3. Multiply or divide from left to right.
4. Add or subtract from left to right.

7. $6 \times 6 \times 6 \times 6 = 6^4$

9. $5 \times 5 \times 5 \times 5 \times 5 \times 5 = 5^6$

11. $8 \times 8 \times 8 \times 8 = 8^4$

13. $9 = 9^1$

15. $2^4 = 2 \times 2 \times 2 \times 2 = 16$

17. $4^3 = 4 \times 4 \times 4 = 64$

19. $6^2 = 6 \times 6 = 36$

21. $10^4 = 10 \times 10 \times 10 \times 10 = 10,000$

23. $1^{17} = 1$

25. $2^6 = 2\times2\times2\times2\times2\times2 = 64$

27. $3^5 = 3\times3\times3\times3\times3 = 243$

29. $15^2 = 15\times15 = 225$

31. $7^3 = 7\times7\times7 = 343$

23. $4^4 = 4\times4\times4\times4 = 256$

35. $9^0 = 1$

37. $25^2 = 25\times25 = 625$

39. $10^6 = 10\times10\times10\times10\times10\times10 = 1,000,000$

41. $13^2 = 13\times13 = 169$

43. $9^1 = 9$

45. $7^4 = 7\times7\times7\times7 = 2401$

47. $2^4 + 1^8 = 16 + 1 = 17$

49. $6^3 + 3^2 = 216 + 9 = 225$

51. $8^3 + 8 = 512 + 8 = 520$

53. $7\times8 - 4 = 56 - 4 = 52$

55. $3\times9 - 10\div2 = 27 - 5 = 22$

57. $$\begin{aligned}48\div2^3 + 4 &= 48\div8 + 4\\ &= 6 + 4\\ &= 10\end{aligned}$$

59. $$\begin{aligned}3\times10^2 - 50 &= 3\times100 - 50\\ &= 300 - 50\\ &= 250\end{aligned}$$

61. $$\begin{aligned}&10^2 + 3\times(8-3)\\ &= 10^2 + 3(5)\\ &= 100 + 3(5)\\ &= 100 + 15\\ &= 115\end{aligned}$$

63. $(400\div20)\div20 = 20\div20 = 1$

65. $950\div(25\div5) = 950\div5 = 190$

67. $$\begin{aligned}(12)(5) - (12+5) &= (12)(5) - 17\\ &= 60 - 17\\ &= 43\end{aligned}$$

69. $$\begin{aligned}3^2 + 4^2\div2^2 &= 9 + 16\div4\\ &= 9 + 4\\ &= 13\end{aligned}$$

71. $$\begin{aligned}(6)(7) - (12-8)\div4 &= (6)(7) - 4\div4\\ &= 42 - 1\\ &= 41\end{aligned}$$

73. $$\begin{aligned}100 - 3^2\times4 &= 100 - 9\times4\\ &= 100 - 36\\ &= 64\end{aligned}$$

75. $5^2 + 2^2 + 3^3 = 25 + 4 + 27 = 56$

77. $$\begin{aligned}72\div9\times3\times1\div2 &= 8\times3\times1\div2\\ &= 24\times1\div2\\ &= 24\div2\\ &= 12\end{aligned}$$

79. $12^2 - 6\times3\times4\times0 = 144 - 0 = 144$

81. $$\begin{aligned}4^2\times6\div3 &= 16\times6\div3\\ &= 96\div3\\ &= 32\end{aligned}$$

83. $16-(27\div 9)\times 4+1$
$=16-3\times 4+1$
$=16-12+1$
$=5$

85. $3+3^2\times 6+4=3+9\times 6+4$
$=3+54+4$
$=61$

87. $32\div 2\times (3-1)^4=32\div 2\times 2^4$
$=32\div 2\times 16$
$=16\times 16$
$=256$

89. $3^2\times 6\div 9+4\times 3$
$=9\times 6\div 9+4\times 3$
$=54\div 9+4\times 3$
$=6+12$
$=18$

91. $6^2+3^4=36+81$
$=117$

93. $1200-2^3(3)\div 6=1200-8(3)\div 6$
$=1200-24\div 6$
$=1200-4$
$=1196$

95. $250\div 5+20-3^2$
$=250\div 5+20-9$
$=50+20-9$
$=61$

97. $250\div (5+20)-3^2$
$=250\div (25)-3^2$
$=250\div 25-9$
$=10-9$
$=1$

99. $2\times 3+(11-5)^2-4\times 5$
$=2\times 3+(6)^2-4\times 5$
$=2\times 3+36-4\times 5$
$=6+36-4\times 5$
$=6+36-20$
$=22$

101. $23(60)(60)+56(60)+4=82,800+3360+4$
$=86,164$ seconds

Cumulative Review

103. a. 3
b. 2,000,000

105. 261,763,002
Two hundred sixty-one million, seven hundred sixty-three thousand, two

1.7 Exercises

1. Locate the rounding place. If the digit to the right of the rounding place is greater than or equal to 5, round up. If the digit to the right of the rounding place is less than 5, round down.

3. $8\underline{3}$ rounds to 80 since 3 is less than 5.

5. $6\underline{5}$ rounds to 70 since 5 is equal to 5.

7. $16\underline{8}$ rounds to 170 since 8 is greater than 5.

9. $743\underline{8}$ rounds to 7440 since 8 is greater than 5.

11. $167\underline{2}$ rounds to 1670 since 2 is less than 5.

13. $2\underline{4}7$ rounds to 200 since 4 is less than 5.

15. $27\underline{8}1$ rounds to 2800 since 8 is greater than 5.

17. $76\underline{9}2$ rounds to 7700 since 9 is greater than 5.

19. $7\underline{6}21$ rounds to 8000 since 6 is greater than 5.

21. 1672 rounds to 2000 since 6 is greater than 5.

23. 27,863 rounds to 28,000 since 8 is greater than 5.

25. 832,400 rounds to 800,000 since 3 is less than 5.

27. 15,169,873 rounds to 15,000,000 stars since 1 is less than 5.

29. a. 163,298 rounds to 163,000 since 2 is less than 5.
b. 163,298 rounds to 163,300 since 9 is greater than 5.

31. a. 3,705,392 rounds to 3,700,000 square miles since 0 is less than 5.
9,596,960 rounds to 9,600,000 square kilometers since 9 is greater than 5.
b. 3,705,392 rounds to 3,710,000 square miles since 1 is equal to 5.
9,596,960 rounds to 9,600,000 square kilometers since 6 is greater than 5.

33. $600 + 300 + 100 = 1000$

35.
$$\begin{array}{r} 40 \\ 70 \\ 100 \\ +\ 20 \\ \hline 230 \end{array}$$

37. $200,000 + 50,000 + 9,000 = 259,000$

39. $600,000 - 100,000 = 500,000$

41. $800,000 - 80,000 = 720,000$

43. $30,000,000 - 20,000,000 = 10,000,000$

45. $60 \times 50 = 3000$

47. $1000 \times 8 = 8,000$

49. $600,000 \times 300 = 180,000,000$

51. $6,000 \div 30 = 200$

53. $200,000 \div 40 = 5,000$

55. $4,000,000 \div 800 = 5000$

57.
$$\begin{array}{r} 400 \\ 500 \\ 900 \\ +\ 200 \\ \hline 2000 \end{array}$$
Incorrect

59.
$$\begin{array}{r} 100,000 \\ 50,000 \\ +\ 40,000 \\ \hline 190,000 \end{array}$$
Incorrect

61.
$$\begin{array}{r} 300,000 \\ -\ 90,000 \\ \hline 210,000 \end{array}$$
Correct

63.
$$\begin{array}{r} 80,000,000 \\ -\ 50,000,000 \\ \hline 30,000,000 \end{array}$$
Incorrect

65.
$$\begin{array}{r} 200 \\ \times\ 20 \\ \hline 4000 \end{array}$$
Incorrect

67.
$$\begin{array}{r} 6000 \\ \times\ 70 \\ \hline 420,000 \end{array}$$
Correct

69. $40\overline{)80,000}$ quotient 2000 Correct

71. $400\overline{)200,000}$ quotient 500

Correct

73. $60 \times 40 = 2400$ square feet

75. $2,000,000 + 3,000,000 + 300,000 = 5,300,000$

77. $300 \times 100 = 30,000$ pizzas

79. $600,000,000 - 400,000,000$
$= 200,000,000$ passengers

81. $590,000 - 270,000 = 320,000$ square miles

83. a. $8,000,000,000 \div 20,000$
$= 400,000$ hours

b. $400,000 \div 20 = 20,000$ days

Cumulative Review

85. $26 \times 3 + 20 \div 4 = 78 + 20 \div 4 = 78 + 5 = 83$

87. $3 \times (16 \div 4) + 8 \times 2 = 3 \times 4 + 8 \times 2 = 12 + 16 = 28$

89.
$$\begin{array}{r} 5489 \\ \times \quad 67 \\ \hline 38423 \\ 32934 \\ \hline 367,763 \end{array}$$

1.8 Exercises

1.
$$\begin{array}{r} 40,300 \\ - \quad 31,500 \\ \hline 8800 \end{array}$$

The repairs will cost \$8800

3.
$$\begin{array}{r} 120 \\ \times \quad 13 \\ \hline 360 \\ 120 \\ \hline 1560 \end{array}$$ bagels

5. $96 \div 16$

$$\begin{array}{r} 6 \\ 16\overline{)96} \\ \underline{96} \\ 0 \end{array}$$

They costs 6 cents per ounce.

7.
$$\begin{array}{r} 64 \\ 13\overline{)832} \\ \underline{78} \\ 52 \\ \underline{52} \\ 0 \end{array}$$

Each pair costs \$64

9. $300 \div 60 = 5$

It will take him 5 hours or 300 minutes.

11.
$$\begin{array}{r} 7356 \\ 3257 \\ 4777 \\ + \quad 4992 \\ \hline 20,382 \end{array}$$

The gross revenue was \$20,382.

13.
$$\begin{array}{r} 24,111 \\ 327 \\ + \quad 793 \\ \hline 25,231 \end{array} \qquad \begin{array}{r} 793 \\ -\ 327 \\ \hline \end{array}$$

466 more volunteers

15. $480 \div 60 = 8$

$100,000 \times 8 = 800,000$ people

17.
$$\begin{array}{r} 15 \\ \times 6 \\ \hline 90 \end{array} \quad \begin{array}{r} 9 \\ \times 8 \\ \hline 72 \end{array} \quad \begin{array}{r} 5 \\ \times 6 \\ \hline 30 \end{array}$$

$90 + 72 + 30 = 192$

She made \$192.

19.
$$\begin{array}{r} 61 \\ 385 \\ 945 \\ 732 \\ +\,144 \\ \hline 2267 \end{array} \qquad \begin{array}{r} 223 \\ 29 \\ 98 \\ +\,435 \\ \hline 785 \end{array} \qquad \begin{array}{r} 2267 \\ -\,785 \\ \hline 1482 \end{array}$$

The balance is $1482.

21.
$$\begin{array}{r} 250 \\ \times\,85 \\ \hline 1250 \\ 2000 \\ \hline 21{,}250 \end{array} \qquad \begin{array}{r} 57 \\ \times\,85 \\ \hline 285 \\ 456 \\ \hline 4845 \end{array} \qquad \begin{array}{r} 21{,}250 \\ -4\,845 \\ \hline 16{,}405 \end{array}$$

Her profit is $16,405.

23.
$$\begin{array}{r} 15{,}276 \\ -\,14{,}926 \\ \hline 350 \end{array}$$

$350 \div 14 = 25$

Her car gets 25 miles per gallon.

25. oaks $3 \times 18 = 54$

maples $2 \times 54 = 108$

pines $7 \times 108 = 756$

birches $= 18$

936 trees

27.
$$\begin{array}{r} 53 \\ 44 \\ +\,21 \\ \hline 118 \end{array}$$ students

29.
$$\begin{array}{r} 174 \\ +\,226 \\ \hline 400 \end{array} \qquad \begin{array}{r} 400 \\ -\,183 \\ \hline 217 \end{array}$$ more students

31.
$$\begin{array}{r} 33{,}867 \\ -19{,}133 \\ \hline \$14{,}734 \end{array}$$ million

or $14,734,000,000

33.
$$\begin{array}{r} 51{,}735 \\ -\,33{,}867 \\ \hline 17{,}868 \end{array} \qquad \begin{array}{r} 51{,}735 \\ +\,17{,}868 \\ \hline 69{,}603 \end{array}$$

Expenditures will be $69,603,000,000

Cumulative Review

35. $7^3 = 7 \times 7 \times 7 = 343$

37.
$$\begin{array}{r} 126 \\ \times\,38 \\ \hline 1008 \\ 378 \\ \hline 4788 \end{array}$$

39.
$$\begin{array}{r} 96 \\ 123 \\ 57 \\ +\,526 \\ \hline 802 \end{array}$$

41. $526{,}195{,}\underline{7}26$ rounds to 526,196,000 because 7 is greater than 5.

Putting Your Skills to Work

1.
$$\begin{array}{r} 529 \\ -\,455 \\ \hline 74 \end{array}$$ wells

2.
$$\begin{array}{r} 455 \\ 529 \\ +\,837 \\ \hline 1821 \end{array}$$ wells

3.
$$\begin{array}{r} 780 \\ -\ 450 \\ \hline 330 \end{array}$$
330,000 people

4.
$$\begin{array}{r} 1130 \\ -\ 780 \\ \hline 350 \end{array}$$
350,000 people

5.
$$\begin{array}{r} 989 \text{ R5} \\ 455\overline{)450,000} \\ \underline{409\ 5} \\ 40\ 50 \\ \underline{36\ 40} \\ 4\ 100 \\ \underline{4\ 095} \\ 5 \end{array}$$
989 people

6.
$$\begin{array}{r} 623 \text{ R433} \\ 529\overline{)330,000} \\ \underline{317\ 4} \\ 12\ 60 \\ \underline{10\ 58} \\ 2\ 020 \\ \underline{1\ 587} \\ 433 \end{array}$$
624 people

Chapter 1 Review Problems

1. Three hundred seventy-six

2. Fifteen thousand eight hundred two

3. One hundred nine thousand, two hundred seventy-six

4. Four hundred twenty-three million, five hundred seventy-six thousand, fifty-five

5. $4364 = 4000 + 300 + 60 + 4$

6. $27,986 = 20,000 + 7000 + 900 + 80 + 6$

7. $42,166,037 = 40,000,000 + 2,000,000 + 100,000 + 60,000 + 6,000 + 30 + 7$

8. $1,305,128 = 1,000,000 + 300,000 + 5000 + 100 + 20 + 8$

9. 924

10. 5302

11. 1,328,828

12. 45,092,651

13.
$$\begin{array}{r} 76 \\ +39 \\ \hline 115 \end{array}$$

14.
$$\begin{array}{r} 148 \\ +152 \\ \hline 300 \end{array}$$

15.
$$\begin{array}{r} 127 \\ +563 \\ \hline 690 \end{array}$$

16.
$$\begin{array}{r} 12 \\ 28 \\ 34 \\ +76 \\ \hline 150 \end{array}$$

17.
$$\begin{array}{r} 122 \\ 61 \\ 9 \\ 84 \\ +123 \\ \hline 400 \end{array}$$

18. $\begin{array}{r} 937 \\ 405 \\ +256 \\ \hline 1598 \end{array}$

19. $\begin{array}{r} 226 \\ 134 \\ +647 \\ \hline 1007 \end{array}$

20. $\begin{array}{r} 28,364 \\ +97,059 \\ \hline 125,423 \end{array}$

21. $\begin{array}{r} 1356 \\ 2892 \\ 561 \\ 89 \\ +9805 \\ \hline 14,703 \end{array}$

22. $\begin{array}{r} 26 \\ 503 \\ 935 \\ 1257 \\ +7861 \\ \hline 10,582 \end{array}$

23. $\begin{array}{r} 36 \\ -\ 19 \\ \hline 17 \end{array}$

24. $\begin{array}{r} 54 \\ -\ 48 \\ \hline 6 \end{array}$

25. $\begin{array}{r} 126 \\ -\ 99 \\ \hline 27 \end{array}$

26. $\begin{array}{r} 543 \\ -\ 372 \\ \hline 171 \end{array}$

27. $\begin{array}{r} 1296 \\ -\ 1137 \\ \hline 159 \end{array}$

28. $\begin{array}{r} 9000 \\ -\ 5833 \\ \hline 3167 \end{array}$

29. $\begin{array}{r} 201,010 \\ -\ 137,864 \\ \hline 63,146 \end{array}$

30. $\begin{array}{r} 101,300 \\ -\ 98,274 \\ \hline 3,026 \end{array}$

31. $\begin{array}{r} 6,325,034 \\ -\ 89,023 \\ \hline 6,236,011 \end{array}$

32. $\begin{array}{r} 5,412,022 \\ -\ 79,031 \\ \hline 5,332,991 \end{array}$

33. $8 \times 1 \times 9 \times 2 = 8 \times 9 \times 2 = 72 \times 2 = 144$

34. $7 \times 6 \times 0 \times 4 = 42 \times 0 \times 4 = 0 \times 4 = 0$

35. $3 \cdot 4 \cdot 2 \cdot 2 \cdot 5 = 12 \cdot 2 \cdot 2 \cdot 5 = 24 \cdot 2 \cdot 5 = 48 \cdot 5 = 240$

36. $1 \cdot 3 \cdot 10 \cdot 5 \cdot 2 = 3 \cdot 10 \cdot 5 \cdot 2 = 30 \cdot 5 \cdot 2 = 150 \cdot 2 = 300$

37. $621 \times 100 = 62,100$

38. $84,312 \times 1000 = 84,312,000$

39. $832 \times 100,000 = 83,200,000$

40. $563 \times 1{,}000{,}000 = 563{,}000{,}000$

41.
$$\begin{array}{r} 58 \\ \times\ 32 \\ \hline 1856 \end{array}$$

42.
$$\begin{array}{r} 73 \\ \times\ 24 \\ \hline 292 \\ 146 \\ \hline 1752 \end{array}$$

43.
$$\begin{array}{r} 150 \\ \times\ 27 \\ \hline 1050 \\ 300 \\ \hline 4050 \end{array}$$

44.
$$\begin{array}{r} 360 \\ \times\ 38 \\ \hline 2880 \\ 1080 \\ \hline 13{,}680 \end{array}$$

45.
$$\begin{array}{r} 709 \\ \times\ 36 \\ \hline 4254 \\ 2127 \\ \hline 25{,}524 \end{array}$$

46.
$$\begin{array}{r} 502 \\ \times\ 48 \\ \hline 4016 \\ 2008 \\ \hline 24{,}096 \end{array}$$

47.
$$\begin{array}{r} 123 \\ \times\ 714 \\ \hline 492 \\ 123 \\ 861 \\ \hline 87{,}822 \end{array}$$

48.
$$\begin{array}{r} 431 \\ \times\ 623 \\ \hline 1293 \\ 862 \\ 2586 \\ \hline 268{,}513 \end{array}$$

49.
$$\begin{array}{r} 1782 \\ \times\ \ 305 \\ \hline 8910 \\ 53460 \\ \hline 543{,}510 \end{array}$$

50.
$$\begin{array}{r} 2057 \\ \times\ \ 124 \\ \hline 8228 \\ 4114 \\ 2057 \\ \hline 255{,}068 \end{array}$$

51.
$$\begin{array}{r} 300 \\ \times\ \ 500 \\ \hline 150{,}000 \end{array}$$

52.
$$\begin{array}{r} 400 \\ \times\ \ 600 \\ \hline 240{,}000 \end{array}$$

53.
$$\begin{array}{r} 1200 \\ \times\ \ 6000 \\ \hline 7{,}200{,}000 \end{array}$$

54.
$$\begin{array}{r} 2500 \\ \times \quad 3000 \\ \hline 7{,}500{,}000 \end{array}$$

55.
$$\begin{array}{r} 100{,}000 \\ \times \quad 20{,}000 \\ \hline 2{,}000{,}000{,}000 \end{array}$$

56.
$$\begin{array}{r} 300{,}000 \\ \times \quad 40{,}000 \\ \hline 12{,}000{,}000{,}000 \end{array}$$

57. $20 \div 10 = 2$

58. $40 \div 8 = 5$

59. $0 \div 8 = 0$

60. $12 \div 1 = 12$

61. $7 \div 1 = 7$

62. $0 \div 5 = 0$

63. $\frac{49}{7} = 7$

64. $\frac{42}{6} = 7$

65. $\frac{5}{0}$ undefined

66. $\frac{24}{6} = 4$

67. $\frac{56}{8} = 7$

68. $\frac{48}{8} = 6$

69.
$$\begin{array}{r} 125 \\ 6\overline{)750} \\ \underline{6} \\ 15 \\ \underline{12} \\ 30 \\ \underline{30} \\ 0 \end{array}$$

70.
$$\begin{array}{r} 125 \\ 7\overline{)875} \\ \underline{2} \\ 17 \\ \underline{14} \\ 35 \\ \underline{35} \\ 0 \end{array}$$

71.
$$\begin{array}{r} 258 \\ 5\overline{)1290} \\ \underline{10} \\ 29 \\ \underline{25} \\ 40 \\ \underline{40} \\ 0 \end{array}$$

72.
$$\begin{array}{r} 309 \\ 4\overline{)1236} \\ \underline{12} \\ 36 \\ \underline{36} \\ 0 \end{array}$$

73. $3\overline{)77,622}$ quotient 25,874
6
17
15
26
24
22
21
12
12
0

74. $8\overline{)24,512}$ quotient 3064
24
51
48
32
32
0

75. $6\overline{)221,748}$ quotient 36,958
18
41
36
57
54
34
30
48
48
0

76. $5\overline{)184,605}$ quotient 36,921
15
34
30
46
45
10
10
5
5
0

77. $8\overline{)127,890}$ quotient 15,986
8
47
40
78
72
69
64
50
48
2
15,986 R 2

78. $7\overline{)250,485}$ quotient 35,783
21
40
35
54
49
58
56
25
21
4
35,783 R 4

79. $\begin{array}{r} 7 \\ 67\overline{)490} \\ \underline{469} \\ 21 \end{array}$

7 R 21

80. $\begin{array}{r} 4 \\ 72\overline{)325} \\ \underline{288} \\ 37 \end{array}$

4 R 37

81. $\begin{array}{r} 31 \\ 21\overline{)666} \\ \underline{63} \\ 36 \\ \underline{21} \\ 15 \end{array}$

31 R 15

82. $\begin{array}{r} 14 \\ 22\overline{)319} \\ \underline{22} \\ 99 \\ \underline{88} \\ 11 \end{array}$

14 R 11

83. $\begin{array}{r} 38 \\ 68\overline{)2614} \\ \underline{204} \\ 574 \\ \underline{544} \\ 30 \end{array}$

38 R 30

84. $\begin{array}{r} 60 \\ 53\overline{)3202} \\ \underline{318} \\ 22 \end{array}$

60 R 22

85. $\begin{array}{r} 95 \\ 45\overline{)4275} \\ \underline{405} \\ 225 \\ \underline{225} \\ 0 \end{array}$

86. $\begin{array}{r} 258 \\ 35\overline{)9030} \\ \underline{70} \\ 203 \\ \underline{175} \\ 280 \\ \underline{280} \\ 0 \end{array}$

87. $\begin{array}{r} 54 \\ 132\overline{)7128} \\ \underline{660} \\ 528 \\ \underline{528} \\ 0 \end{array}$

88. $\begin{array}{r} 19 \\ 204\overline{)3876} \\ \underline{204} \\ 1836 \\ \underline{1836} \\ 0 \end{array}$

89. $13\times 13 = 13^2$

90. $21\times 21\times 21 = 21^3$

91. $8\times 8\times 8\times 8\times 8 = 8^5$

92. $10\times 10\times 10\times 10\times 10\times 10 = 10^6$

93. $2^6 = 2\times 2\times 2\times 2\times 2\times 2 = 64$

94. $3^4 = 3\times 3\times 3\times 3 = 81$

95. $2^7 = 2\times2\times2\times2\times2\times2\times2 = 128$

96. $5^3 = 5\times5\times5 = 125$

97. $7^2 = 7\times7 = 49$

98. $9^2 = 9\times9 = 81$

99. $6^3 = 6\times6\times6 = 216$

100. $4^3 = 4\times4\times4 = 64$

101. $7+2\times3-5 = 7+6-5$
$= 13-5$
$= 8$

102. $6\times2-4+3 = 12-4+3$
$= 8+3$
$= 11$

103. $2^5+4-(5+3^2) = 32+4-(5+9)$
$= 32+4-14$
$= 36-14$
$= 22$

104. $4^3+20\div(2+2^3) = 64+20\div(2+8)$
$= 64+20\div10$
$= 64+2$
$= 66$

105. $3^3\times4-6\div6 = 27\times4-6\div6$
$= 108-1$
$= 107$

106. $20\div20+5^3\times3 = 20\div20+25\times3$
$= 1+75$
$= 76$

107. $2^3\times5\div8+3\times4$
$= 8\times5\div8+3\times4$
$= 40\div8+12$
$= 5+12$
$= 17$

108. $2^3+4\times5-32\div(1+3)^2$
$= 2^3+4\times5-32\div4^2$
$= 8+4\times5-32\div16$
$= 8+20-2$
$= 26$

109. $6\times3+3\times5^2-63\div(5-2)^2$
$= 6\times3+3\times5^2-63\div3^2$
$= 6\times3+3\times25-63\div9$
$= 18+75-7$
$= 86$

110. 127<u>5</u> rounds to 1280.

111. 5<u>8</u>95 rounds to 5900.

112. 15,30<u>5</u> rounds to 15,310.

113. 42,64<u>4</u> rounds to 42,640.

114. 12,3<u>5</u>0 rounds to 12,000.

115. 22,<u>9</u>86 rounds to 23,000.

116. 675,<u>8</u>00 rounds to 676,000.

117. 202,<u>4</u>98 rounds to 202,000.

118. 4,<u>6</u>49,320 rounds to 4,600,000.

119. 9,99<u>5</u>,312 rounds to 10,000,000.

120.
$$\begin{array}{r} 600 \\ 600 \\ 900 \\ +\ 900 \\ \hline 3000 \end{array}$$

121.
$$\begin{array}{r} 30,000 \\ 7,000 \\ +\ 70,000 \\ \hline 107,000 \end{array}$$

122. $\begin{array}{r} 4,000,000 \\ -\ 3,000,000 \\ \hline 1,000,000 \end{array}$

123. $\begin{array}{r} 30,000 \\ -\ 20,000 \\ \hline 10,000 \end{array}$

124. $\begin{array}{r} 1000 \\ \times\ \ \ 6000 \\ \hline 6,000,000 \end{array}$

125. $\begin{array}{r} 3,000,000 \\ \times\ \ \ \ \ \ 9000 \\ \hline 2,700,000,000 \end{array}$

126. $\begin{array}{r} 4,000 \\ 20\overline{)80,000} \\ \underline{80} \\ 0 \end{array}$

127. $\begin{array}{r} 20,000 \\ 400\overline{)8,000,000} \end{array}$

128. $\begin{array}{r} 18 \\ \times\ 12 \\ \hline 36 \\ 18 \\ \hline 216 \text{ cans} \end{array}$

129. $\begin{array}{r} 25 \\ \times\ \ 7 \\ \hline 175\ \text{ words} \end{array}$

130. $\begin{array}{r} 2462 \\ 1997 \\ +\ 2561 \\ \hline 7020 \text{ people} \end{array}$

131. $\begin{array}{r} 26,300 \\ 14,520 \\ +\ 18,650 \\ \hline \$59,470 \end{array}$

132. $\begin{array}{r} 14,630 \\ -\ 4,329 \\ \hline 10,301 \end{array}$

10,301 feet between them.

133. $\begin{array}{r} 11,658 \\ -\ 4,630 \\ \hline \$7,028 \end{array}$

134. $\begin{array}{r} 1356 \\ 24\overline{)32,544} \\ \underline{24} \\ 85 \\ \underline{72} \\ 134 \\ \underline{120} \\ 144 \\ \underline{144} \\ 0 \end{array}$

Cost per passenger was $1356.

135. $\begin{array}{r} 74 \\ 112\overline{)8288} \\ \underline{784} \\ 448 \\ \underline{448} \\ 0 \end{array}$

$74 per bed.

136. Deposits Checks

$$\begin{array}{r} 24 \\ 105 \\ 36 \\ +\ 177 \\ \hline 342 \end{array} \qquad \begin{array}{r} 18 \\ 145 \\ 250 \\ +\ 461 \\ \hline 874 \end{array}$$

$810+342-874=278$

Her balance will be \$278.

137.

$$\begin{array}{r} 56,720 \\ -\ 56,320 \\ \hline 400 \end{array}$$

400 miles

$$\begin{array}{r} 25 \\ 16\overline{)400} \\ \underline{32} \\ 80 \\ \underline{80} \\ 0 \end{array}$$

He got 25 miles per gallon.

138. $3\times 279+4\times 61+2\times 1980$

$=837+244+3960$

$=5041$

The total price was \$5041.

139.

$$\begin{array}{r} 15 \\ \times\ 65 \\ \hline 975 \end{array} \quad \begin{array}{r} 60 \\ \times 12 \\ \hline 720 \end{array} \quad \begin{array}{r} 42 \\ \times 8 \\ \hline 336 \end{array} \quad \begin{array}{r} 975 \\ 720 \\ +336 \\ \hline 2031 \end{array}$$

The total price is \$2031.

140.

$$\begin{array}{r} 55,000,000 \\ -\ 14,500,000 \\ \hline 40,500,000 \end{array}$$

The difference is 40,500,000 tons.

141.

$$\begin{array}{r} 55,000,000 \\ -\ 33,600,000 \\ \hline 21,400,000 \end{array}$$

21,400,000 tons between 1990 and 1995

142.

$$\begin{array}{r} 63,500 \\ -\ 33,600 \\ \hline 29,900 \end{array} \qquad \begin{array}{r} 29,900 \\ +\ 63,500 \\ \hline 93,400 \end{array}$$

93,400,000 tons in 2010.

143.

$$\begin{array}{r} 205 \\ 36 \\ 1983 \\ +\ 60 \\ \hline 2284 \end{array}$$

144.

$$\begin{array}{r} 56,793 \\ -\ 48,926 \\ \hline 7,867 \end{array}$$

145.

$$\begin{array}{r} 396 \\ \times\ \ 28 \\ \hline 3168 \\ \underline{792\ } \\ 11,088 \end{array}$$

146.

$$\begin{array}{r} 129 \\ 37\overline{)4773} \\ \underline{37\ \ } \\ 107\ \\ \underline{74\ } \\ 333 \\ \underline{333} \\ 0 \end{array}$$

147. $4\times 12-(12+9)+2^3\div 4$

$=4\times 12-21+2^3\div 4$

$=4\times 12-21+8\div 4$

$=48-21+2$

$=29$

148.

$$\begin{array}{r} 699 \\ \times\ \ 3 \\ \hline 2097 \end{array} \quad \begin{array}{r} 78 \\ \times\ 2 \\ \hline 156 \end{array} \quad \begin{array}{r} 2097 \\ \times 156 \\ \hline 2253 \end{array} \quad \begin{array}{r} 3000 \\ -2253 \\ \hline 747 \end{array}$$

\$747 in his account.

149. a.
$$\begin{array}{r} 22 \\ \times 15 \\ \hline 110 \\ 22 \\ \hline 330 \end{array}$$
330 square feet

b. $2(22)+2(15)=44+30=74$ feet

How Am I Doing? Chapter 1 Test

1. 44,007,635 = Forty-four million, seven thousand, six hundred thirty-five

2. $26,859 = 20,000 + 6000 + 800 + 50 + 9$

3. Three million, five hundred eighty-one thousand, seventy-six = 3,581,076

4.
$$\begin{array}{r} 189 \\ 26 \\ 12 \\ 528 \\ +\ 76 \\ \hline 831 \end{array}$$

5.
$$\begin{array}{r} 763 \\ 220 \\ +\ 508 \\ \hline 1491 \end{array}$$

6.
$$\begin{array}{r} 135,484 \\ 2,376 \\ 81,004 \\ +\ 100,113 \\ \hline 318,977 \end{array}$$

7.
$$\begin{array}{r} 8961 \\ -\ 894 \\ \hline 8067 \end{array}$$

8.
$$\begin{array}{r} 501,760 \\ -\ 328,902 \\ \hline 172,858 \end{array}$$

9.
$$\begin{array}{r} 18,400,100 \\ -\ 13,174,332 \\ \hline 5,225,768 \end{array}$$

10. $1\times6\times9\times7$
$= 6\times9\times7$
$= 54\times7$
$= 378$

11.
$$\begin{array}{r} 45 \\ \times\ 96 \\ \hline 270 \\ 405 \\ \hline 4320 \end{array}$$

12.
$$\begin{array}{r} 326 \\ \times\ 592 \\ \hline 652 \\ 2934 \\ 1630 \\ \hline 192,992 \end{array}$$

13.
$$\begin{array}{r} 18,491 \\ \times\ 7 \\ \hline 129,437 \end{array}$$

14. $5\overline{)15,071}$ quotient 3014

15
0
0
7
5
21
20
1

3014 R 1

15. $6\overline{)14,148}$ quotient 2358

12
21
18
34
30
48
48
0

16. $37\overline{)13,024}$ quotient 352

111
192
185
74
74
0
20
1

17. $14\times14\times14=14^3$

18. $2^6=2\times2\times2\times2\times2\times2=64$

19. $5+6^2-2\times(9-6)^2$

$=5+6^2-2\times3^2$

$=5+36-2\times9$

$=5+36-18$

$=41-18$

$=23$

20. $2^4+3^3+28\div4$

$=16+27+28\div4$

$=16+27+7$

$=50$

21. $4\times6+3^3\times2+23\div23$

$=4\times6+27\times2+23\div23$

$=24+54+1$

$=78+1$

$=79$

22. $94,\underline{7}68$ rounds to 94,800 since 6 is greater than 5.

23. $6,46\underline{2},431$ rounds to 6,460,000 since 2 is less than 5.

24. $5,\underline{2}78,963$ rounds to 5,300,000 since 7 is greater than 5.

25. $5,000,000\times30,000=150,000,000,000$

26. $1000+3000+4000+8000=16,000$

27. $15\overline{)32,220}$ quotient 2148

30
22
15
72
60
120
120
0

Each person paid $2148.

28.
$$\begin{array}{r} 602 \\ -135 \\ \hline 467 \end{array}$$
467 feet

29. $3\times2+1\times45+2\times21+2\times17$
$= 6+45+42+34$
$= 127$
His total bill was $127.

30.
$$\begin{array}{r} 31 \\ 902 \\ +\ 399 \\ \hline \$1332 \end{array} \qquad \begin{array}{r} 885 \\ 103 \\ 26 \\ 17 \\ +\ 9 \\ \hline \$1040 \end{array}$$
Balance is \$1332 − \$1040 = \$292

31.
$$\begin{array}{r} 6800 \\ \times\ \ 110 \\ \hline 0000 \\ 6800 \\ 6800 \\ \hline 748,000 \end{array}$$
Area of runway is 748,000 square feet.

32. Perimeter is $2\times8+2\times15 = 16+30$
$= 46$ feet

Chapter 2

2.1 Exercises

1. fraction

3. denominator

5. N: 3
D: 5

7. N: 7
D: 8

9. N: 1
D: 17

11. $\frac{1}{3}$

13. $\frac{7}{9}$

15. $\frac{3}{4}$

17. $\frac{3}{7}$

19. $\frac{2}{5}$

21. $\frac{7}{10}$

23. $\frac{5}{8}$

25. $\frac{4}{7}$

27. $\frac{7}{8}$

29. $\frac{2}{5}$

31. $\frac{1}{5}$ 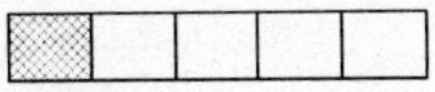

33. $\frac{3}{8}$ 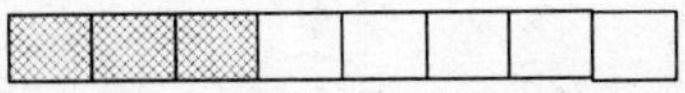

35. $\frac{7}{10}$ 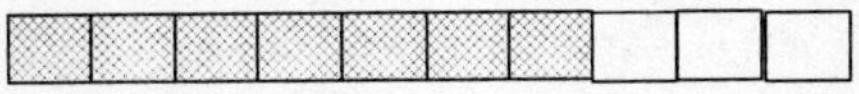

37. $\frac{\text{silver bells}}{\text{wreaths}} = \frac{31}{95}$

39. $\frac{\text{weekend earnings}}{\text{jukebox price}} = \frac{209}{750}$

41. $\frac{\text{roast}}{\text{total}} = \frac{89}{122+89} = \frac{89}{211}$

43. $\frac{\text{rowing}}{\text{total}} = \frac{7}{9+7+13} = \frac{7}{29}$

45. $\frac{\text{ribs or beans}}{\text{total bowls}} = \frac{5+4}{2+3+4+5} = \frac{9}{14}$

47. a. $\frac{50+40}{94+101} = \frac{90}{195}$

b. $\frac{3+19}{94+101} = \frac{22}{195}$

49. $\frac{0}{6}$ is the amount of money each of 6 business owners get if the business has a profit of $0.

Cumulative Review

51.
$$\begin{array}{r} 18 \\ 27 \\ 34 \\ 16 \\ 125 \\ +\ \ 21 \\ \hline 241 \end{array}$$

53.
$$\begin{array}{r} 3178 \\ \times\ \ 46 \\ \hline 19068 \\ 12712\ \\ \hline 146,188 \end{array}$$

55.
$$\begin{array}{r} 282 \\ 866 \\ 42 \\ 317 \\ 102 \\ 99 \\ +\ 115 \\ \hline 1823 \end{array}$$

$$\begin{array}{r} 2004 \\ -\ 1823 \\ \hline 181 \end{array}$$
181 reference books

2.2 Exercises

1. 11, 19, 41, 5

3. composite number

5. $56 = 2\times2\times2\times7$

7. $15 = 3\times5$

9. $35 = 5\times7$

11. $49 = 7\times7 = 7^2$

13. $64 = 8\times8$
$= 2\times2\times2\times2\times2\times2$
$= 2^6$

15. $55 = 5\times11$

17. $63 = 7\times9$
$= 7\times3\times3$
$= 7\times3^2$

19. $75 = 3\times25$
$= 3\times5\times5$
$= 3\times5^2$

21. $54 = 6\times9$
$= 2\times3\times3\times3$
$= 2\times3^3$

23. $120 = 10\times12$
$= 2\times5\times2\times2\times3$
$= 2^3\times3\times5$

25. $184 = 8\times23$
$= 2\times2\times2\times23$
$= 2^3\times23$

27. Prime

29. $57 = 3\times19$

31. Prime

33. $62 = 2\times31$

35. Prime

37. Prime

39. $121 = 11\times11 = 11^2$

41. $129 = 3\times43$

43. $\frac{18}{27} = \frac{18 \div 8}{27 \div 8}$
$= \frac{2}{3}$

45. $\frac{32}{48} = \frac{32 \div 16}{48 \div 16}$
$= \frac{2}{3}$

47. $\frac{30}{48} = \frac{30 \div 6}{48 \div 6}$
$= \frac{5}{8}$

49. $\frac{210}{310} = \frac{210 \div 10}{310 \div 10} = \frac{21}{31}$

51. $\frac{3}{15} = \frac{3 \times 1}{3 \times 5}$
$= \frac{1}{5}$

53. $\frac{66}{88} = \frac{3 \times 2 \times 11}{2 \times 2 \times 2 \times 11}$
$= \frac{3}{4}$

55. $\frac{30}{45} = \frac{2 \times 3 \times 5}{3 \times 3 \times 5}$
$= \frac{2}{3}$

57. $\frac{27}{45} = \frac{3 \times 3 \times 3}{3 \times 3 \times 5}$
$= \frac{3}{5}$

59. $\frac{33}{36} = \frac{3 \times 11}{3 \times 12} = \frac{11}{12}$

61. $\frac{63}{108} = \frac{3 \times 3 \times 7}{2 \times 2 \times 3 \times 3 \times 3} = \frac{7}{12}$

63. $\frac{88}{121} = \frac{11 \times 8}{11 \times 11}$
$= \frac{8}{11}$

65. $\frac{150}{1200} = \frac{3 \times 50}{4 \times 50}$
$= \frac{3}{4}$

67. $\frac{220}{260} = \frac{11 \times 20}{13 \times 20}$
$= \frac{11}{13}$

69. $\frac{3}{11} \stackrel{?}{=} \frac{9}{33}$
$3 \times 33 \stackrel{?}{=} 11 \times 19$
$99 = 99$
Yes

71. $\frac{12}{40} \stackrel{?}{=} \frac{3}{13}$
$12 \times 13 \stackrel{?}{=} 40 \times 3$
$156 \neq 120$
No

73. $\frac{23}{27} \stackrel{?}{=} \frac{92}{107}$
$23 \times 107 \stackrel{?}{=} 27 \times 92$
$2461 \neq 2484$
No

75. $\frac{23}{57} \stackrel{?}{=} \frac{45}{95}$
$27 \times 95 \stackrel{?}{=} 57 \times 45$
$2565 = 2565$
Yes

77. $\frac{65}{70} \stackrel{?}{=} \frac{13}{14}$
$65 \times 14 \stackrel{?}{=} 70 \times 13$
$910 = 910$
Yes

79.
$$\begin{array}{r} 128 \\ -\ 32 \\ \hline 96 \end{array}$$

$$\frac{96}{128} = \frac{16 \times 6}{16 \times 8} = \frac{6}{8} = \frac{3}{4}$$

81. $\frac{95-15}{95} = \frac{80}{95} = \frac{16 \times 5}{19 \times 5}$

$= \frac{16}{19}$ passed

83. $\frac{5{,}000}{17{,}500} = \frac{2500 \times 2}{2500 \times 7} = \frac{2}{7}$

85. Total student body is:

$1100 + 1700 + 900 + 500 + 300 = 4500$

Short cummute is:

$\frac{1700}{4500} = \frac{17 \times 100}{45 \times 100} = \frac{17}{45}$

87. $\frac{500+300}{4500} = \frac{800}{4500}$

$= \frac{8 \times 100}{45 \times 100}$

$= \frac{8}{45}$

Cumulative Review

89.
$$\begin{array}{r} 386 \\ \times\ 425 \\ \hline 1930 \\ 772 \\ 1544 \\ \hline 164{,}050 \end{array}$$

91.
$$\begin{array}{r} 3200 \\ \times\ 300 \\ \hline 960{,}000 \end{array}$$

2.3 Exercises

1. a. Multiply the whole number by the denominator of the fraction.

b. Add the numerator of the fraction to the product formed in step (a).

c. Write the sum found in step (b) over the denominator of the fraction.

3. $4\frac{2}{3} = \frac{3 \times 4 + 2}{3}$

$= \frac{14}{3}$

5. $2\frac{3}{7} = \frac{7 \times 2 + 3}{7}$

$= \frac{17}{7}$

7. $9\frac{2}{9} = \frac{9 \times 9 + 2}{9}$

$= \frac{83}{9}$

9. $10\frac{2}{3} = \frac{3 \times 10 + 2}{3}$

$= \frac{32}{3}$

11. $21\frac{2}{3} = \frac{3 \times 21 + 2}{3}$

$= \frac{65}{3}$

13. $9\frac{1}{6} = \frac{6 \times 9 + 1}{6}$

$= \frac{55}{6}$

15. $20\frac{1}{6} = \frac{6 \times 20 + 1}{6}$

$= \frac{121}{6}$

17. $10\frac{11}{12} = \frac{12 \times 10 + 11}{12}$
$= \frac{131}{12}$

19. $7\frac{9}{10} = \frac{10 \times 7 + 9}{10}$
$= \frac{79}{10}$

21. $8\frac{1}{25} = \frac{25 \times 8 + 1}{25}$
$= \frac{201}{25}$

23. $5\frac{5}{12} = \frac{12 \times 5 + 5}{12}$
$= \frac{65}{12}$

25. $164\frac{2}{3} = \frac{3 \times 164 + 2}{3}$
$= \frac{494}{3}$

27. $8\frac{11}{15} = \frac{15 \times 8 + 11}{15}$
$= \frac{131}{15}$

29. $5\frac{13}{25} = \frac{25 \times 5 + 13}{25}$
$= \frac{138}{25}$

31.
$$\begin{array}{r} 1 \\ 3\overline{)4} \\ \underline{3} \\ 1 \end{array}$$
$\frac{4}{3} = 1\frac{1}{3}$

33.
$$\begin{array}{r} 2 \\ 4\overline{)11} \\ \underline{8} \\ 3 \end{array}$$
$\frac{11}{4} = 2\frac{3}{4}$

35.
$$\begin{array}{r} 2 \\ 6\overline{)15} \\ \underline{12} \\ 3 \end{array}$$
$\frac{15}{6} = 2\frac{3}{6} = 2\frac{1}{2}$

37.
$$\begin{array}{r} 3 \\ 8\overline{)27} \\ \underline{24} \end{array}$$
$\frac{27}{8} = 3\frac{3}{8}$

39.
$$\begin{array}{r} 5 \\ 12\overline{)60} \\ \underline{60} \\ 0 \end{array}$$
$\frac{60}{12} = 5$

41.
$$\begin{array}{r} 9 \\ 9\overline{)86} \\ \underline{81} \\ 5 \end{array}$$
$\frac{86}{9} = 9\frac{5}{9}$

43.
$$\begin{array}{r} 23 \\ 3\overline{)70} \\ \underline{6} \\ 10 \\ \underline{9} \\ 1 \end{array}$$
$\frac{70}{3} = 23\frac{1}{3}$

45. $\begin{array}{r} 3 \\ 16\overline{)51} \\ \underline{48} \\ 3 \end{array}$

$\frac{51}{16} = 3\frac{3}{16}$

47. $\begin{array}{r} 9 \\ 3\overline{)28} \\ \underline{27} \\ 1 \end{array}$

$\frac{28}{3} = 9\frac{1}{3}$

49. $\begin{array}{r} 17 \\ 2\overline{)35} \\ \underline{2} \\ 15 \\ \underline{14} \\ 1 \end{array}$

$\frac{35}{2} = 17\frac{1}{2}$

51. $\begin{array}{r} 13 \\ 7\overline{)91} \\ \underline{7} \\ 21 \\ \underline{21} \\ 0 \end{array}$

$\frac{91}{7} = 13$

53. $\begin{array}{r} 14 \\ 15\overline{)210} \\ \underline{15} \\ 60 \\ \underline{60} \\ 0 \end{array}$

$\frac{210}{15} = 14$

55. $\begin{array}{r} 6 \\ 17\overline{)102} \\ \underline{102} \\ 0 \end{array}$

$\frac{102}{17} = 6$

57. $\begin{array}{r} 36 \\ 11\overline{)403} \\ \underline{33} \\ 73 \\ \underline{66} \\ 7 \end{array}$

$\frac{403}{11} = 36\frac{7}{11}$

59. $2\frac{9}{12} = 2\frac{3\times3}{3\times4} = 2\frac{3}{4}$

61. $4\frac{11}{66} = 4\frac{11\times1}{11\times6} = 4\frac{1}{6}$

63. $15\frac{18}{72} = 15\frac{1}{4}$

65. $\frac{24}{6} = \frac{6\times4}{6\times1} = 4$

67. $\frac{36}{15} = \frac{12\times3}{5\times3} = \frac{12}{5}$

69. $\begin{array}{r} 6 \\ 14\overline{)91} \\ \underline{84} \\ 7 \end{array}$

$\frac{91}{14} = 6\frac{7}{14} = 6\frac{1}{2}$

71. $126\overline{)340}$ with quotient 2

$\underline{252}$

88

$$\frac{340}{126} = 2\frac{88}{126} = 2\frac{44 \times 2}{63 \times 2} = 2\frac{44}{63}$$

73. $280\overline{)580}$ with quotient 2

$\underline{560}$

20

$$\frac{580}{280} = 2\frac{20}{280} = 2\frac{1}{14}$$

75. $296\overline{)508}$ with quotient 1

$\underline{296}$

212

$$\frac{508}{296} = 1\frac{212}{296} = 1\frac{53 \times 4}{74 \times 4} = 1\frac{53}{74}$$

77. $$360\frac{2}{3} = \frac{3 \times 360 + 2}{3} = \frac{1082}{3} \text{ yards}$$

79. $3\overline{)151}$ with quotient 50

$\underline{15}$

1

$\underline{0}$

1

$$\frac{151}{3} = 50\frac{1}{3} \text{ acres}$$

81. $8\overline{)1131}$ with quotient 141

$\underline{8}$

33

$\underline{32}$

11

$\underline{8}$

3

$$\frac{1131}{8} = 141\frac{3}{8} \text{ pounds}$$

83. No; 101 is prime and is not a factor of 5687.

Cumulative Review

85.
$$\begin{array}{r} 1,398,210 \\ -\ 1,137,963 \\ \hline 260,247 \end{array}$$

87. $300{,}000 \div 1000 = 300$

2.4 Exercises

1. $\frac{3}{5} \times \frac{7}{11} = \frac{21}{55}$

3. $\frac{3}{4} \times \frac{5}{13} = \frac{15}{52}$

5. $\frac{6}{5} \times \frac{10}{12} = \frac{1}{1} \times \frac{2}{2} = 1$

7. $\frac{7}{36} \times \frac{30}{9} = \frac{7}{6} \times \frac{5}{9} = \frac{35}{54}$

9. $\frac{15}{28} \times \frac{7}{9} = \frac{5}{4} \times \frac{1}{3} = \frac{5}{12}$

11. $\frac{9}{10} \times \frac{33}{12} = \frac{3}{2} \times \frac{7}{4} = \frac{21}{8} = 2\frac{5}{8}$

13. $8 \times \frac{3}{7} = \frac{8}{1} \times \frac{3}{7} = \frac{24}{7}$ or $3\frac{3}{7}$

15. $\frac{15}{12}\times 8=\frac{5}{3\times 4}\times\frac{2\times 4}{1}=\frac{10}{3}=3\frac{1}{3}$

17. $\frac{4}{9}\times\frac{3}{7}\times\frac{7}{8}=\frac{1}{3}\times\frac{1}{1}\times\frac{1}{2}=\frac{1}{6}$

19. $\frac{5}{4}\times\frac{9}{10}\times\frac{8}{3}=\frac{5}{4}\times\frac{3\times 3}{2\times 5}=\frac{2\times 4}{3}=3$

21. $2\frac{3}{4}\times\frac{8}{9}=\frac{11}{4}\times\frac{8}{9}=\frac{22}{9}=2\frac{4}{9}$

23. $10\times 3\frac{1}{10}=\frac{10}{1}\times\frac{31}{10}=31$

25. $1\frac{3}{16}\times 0=\frac{19}{16}\times 0=0$

27. $3\frac{7}{8}\times 1=3\frac{7}{8}$

29. $1\frac{1}{4}\times 3\frac{2}{3}=\frac{5}{4}\times\frac{11}{3}=\frac{55}{12}=4\frac{7}{12}$

31. $2\frac{3}{10}\times\frac{3}{5}=\frac{23}{10}\times\frac{3}{5}=\frac{69}{50}=1\frac{19}{50}$

33. $4\frac{1}{5}\times 12\frac{2}{9}=\frac{21}{8}\times\frac{110}{9}=\frac{154}{3}=51\frac{1}{3}$

35. $6\frac{2}{5}\times\frac{1}{4}=\frac{32}{5}\times\frac{1}{4}=\frac{8}{5}=1\frac{3}{5}$

37. $\frac{11}{15}\times\frac{35}{33}=\frac{11}{3\times 5}\times\frac{5\times 7}{3\times 11}=\frac{7}{9}$

39. $3\frac{1}{4}\times 4\frac{2}{3}=\frac{13}{4}\times\frac{14}{3}$
$=\frac{13}{2\times 2}\times\frac{2\times 7}{3}$
$=\frac{91}{6}$ or $15\frac{1}{6}$

41. $\frac{2}{7}\cdot x=\frac{18}{35}$
$\frac{2\cdot 9}{7\cdot 5}=\frac{18}{35}$
$x=\frac{9}{5}$

43. $\frac{7}{13}\cdot x=\frac{56}{117}$
$\frac{7\cdot 8}{13\cdot 9}=\frac{56}{117}$
$x=\frac{8}{9}$

45. $8\frac{3}{4}\times 4\frac{1}{3}=\frac{4\times 8+3}{4}\times\frac{3\times 4+1}{3}$
$=\frac{35}{4}\times\frac{13}{3}$
$=\frac{455}{12}$
$=37\frac{11}{12}$ square miles

47. $360\times 4\frac{1}{3}=\frac{360}{1}\times\frac{13}{3}$
$=120\times 13$
$=1560$ miles

49. $90\frac{1}{2}\times 18=\frac{181}{2}\times\frac{18}{1}$
$=181\times 9$
$=1629$ grams

51. $\frac{1}{18}\times 396=\frac{1}{18}\times\frac{18\times 22}{1}$
$=22$ students

53. $12{,}064\times\frac{1}{32}=\frac{12{,}064}{32}$
$=377$ companies

55. $4\frac{1}{4}\times 1\frac{1}{3}=\frac{17}{4}\times\frac{4}{3}=\frac{17}{3}$

$\frac{17}{3}\times\frac{1}{3}=\frac{17}{9}=1\frac{8}{9}$ miles

57. a. $\frac{1}{2}\times\frac{2}{3}=\frac{1}{3}$ of the garden

b. $\frac{1}{3}\times 120=\frac{1}{3}\times\frac{3\times 40}{1}=40\text{ ft}^2$

59. The step of dividing the numerator and denominator by the same number allows us to work with smaller numbers when we do the multiplication. Also, this allows us to avoid the step of having to simplify the fraction in the final answer.

Cumulative Review

61.
$$\begin{array}{r} 529 \\ 31\overline{)16{,}399} \\ \underline{155} \\ 89 \\ \underline{62} \\ 279 \\ \underline{279} \\ 0 \end{array}$$

529 cars

63.
$$\begin{array}{r} 240 \\ \times\ \ 21 \\ \hline 240 \\ \underline{480} \\ 5040 \end{array}$$

5040 miles

2.5 Exercises

1. Think of a simple problem like $3\div\frac{1}{2}$. One way to think of it is how many $\frac{1}{2}$'s can be placed in 3? For example, how many $\frac{1}{2}$ pound rocks could be put in a bag that holds 3 pounds of rocks? The answer is 6. If we inverted the first fraction by mistake, we would have $\frac{1}{3}\times\frac{1}{2}=\frac{1}{6}$. We know that is wrong since there are obviously several $\frac{1}{2}$ pound rocks in a bag that holds 3 pounds of rocks. The answer $\frac{1}{6}$ would make no sense.

3. $\frac{7}{8}\div\frac{2}{3}=\frac{7}{8}\times\frac{3}{2}=\frac{21}{16}=1\frac{5}{16}$

5. $\frac{2}{3}\div\frac{4}{27}=\frac{2}{3}\times\frac{27}{4}$

$=\frac{9}{2}=4\frac{1}{2}$

7. $\frac{5}{9}\div\frac{10}{27}=\frac{5}{9}\times\frac{27}{10}$

$=\frac{3}{2}=1\frac{1}{2}$

9. $\frac{2}{9}\div\frac{1}{6}=\frac{2}{9}\times\frac{6}{1}$

$=\frac{4}{3}=1\frac{1}{3}$

11. $\frac{4}{15}\div\frac{4}{15}=\frac{4}{15}\times\frac{15}{4}=1$

13. $\frac{3}{7}\div\frac{7}{3}=\frac{3}{7}\times\frac{3}{7}=\frac{9}{49}$

15. $\frac{4}{5}\div 1=\frac{4}{5}\times\frac{1}{1}=\frac{4}{5}$

17. $\frac{3}{11} \div 4 = \frac{3}{11} \times \frac{1}{4} = \frac{3}{44}$

19. $1 \div \frac{7}{27} = \frac{1}{1} \times \frac{27}{7}$
$= \frac{27}{7} \text{ or } 3\frac{6}{7}$

21. $0 \div \frac{3}{17} = 0 \times \frac{17}{3} = 0$

23. undefined

25. $8 \div \frac{4}{5} = \frac{8}{1} \times \frac{5}{4} = \frac{10}{1} = 10$

27. $\frac{7}{8} \div 4 = \frac{7}{8} \times \frac{1}{4} = \frac{7}{32}$

29. $\frac{9}{16} \div \frac{3}{4} = \frac{9}{16} \times \frac{4}{3} = \frac{3}{4}$

31. $3\frac{1}{4} \div 2\frac{1}{4} = \frac{13}{4} \div \frac{4}{9}$
$= \frac{13}{9} \text{ or } 1\frac{4}{9}$

33. $6\frac{2}{5} \div 3\frac{1}{5} = \frac{32}{5} \times \frac{5}{16} = 2$

35. $6000 \div \frac{6}{5} = \frac{6000}{1} \times \frac{5}{6} = 5000$

37. $\frac{\frac{4}{5}}{200} = \frac{4}{5} \times \frac{1}{200} = \frac{1}{250}$

39. $\frac{\frac{5}{8}}{\frac{25}{7}} = \frac{5}{8} \times \frac{7}{25} = \frac{7}{40}$

41. $3\frac{1}{5} \div \frac{3}{10} = \frac{16}{5} \times \frac{10}{3} = \frac{32}{3} = 10\frac{2}{3}$

43. $2\frac{1}{3} \div 6 = \frac{7}{3} \times \frac{1}{6} = \frac{7}{18}$

45. $5\frac{1}{4} \div 2\frac{5}{8} = \frac{21}{4} \div \frac{21}{8}$
$= \frac{21}{4} \times \frac{8}{21}$
$= 2$

47. $5 \div 1\frac{1}{4} = \frac{5}{1} \div \frac{5}{4}$
$= \frac{5}{1} \times \frac{4}{5}$
$= 4$

49. $12\frac{1}{2} \div 5\frac{5}{6} = \frac{25}{2} \div \frac{35}{6}$
$= \frac{25}{2} \times \frac{6}{35}$
$= \frac{15}{7} = 2\frac{1}{7}$

51. $8\frac{1}{4} \div 2\frac{3}{4} = \frac{33}{4} \div \frac{11}{4}$
$= \frac{33}{4} \times \frac{4}{11}$
$= 3$

53. $3\frac{1}{2} \times \frac{9}{16} = \frac{7}{2} \times \frac{9}{16} = \frac{63}{32} = 1\frac{31}{32}$

55. $3\frac{3}{4} \div 9 = \frac{15}{4} \times \frac{1}{9} = \frac{5}{12}$

57. $\frac{5}{3\frac{1}{6}} = \frac{5}{1} \times \frac{6}{19} = \frac{30}{19} \text{ or } 1\frac{11}{19}$

59. $\frac{0}{4\frac{3}{8}} = \frac{0}{\frac{35}{8}} = 0 \times \frac{8}{35} = 0$

61. $\frac{\frac{7}{12}}{3\frac{2}{3}} = \frac{7}{12} \div \frac{11}{3} = \frac{7}{12} \times \frac{3}{11} = \frac{7}{44}$

63. $3\frac{3}{5} \times 2\frac{1}{3} = \frac{18}{5} \times \frac{7}{3}$
$= \frac{42}{5} \text{ or } 8\frac{2}{5}$

65. $x \div \frac{4}{3} = \frac{21}{20}$

$x \times \frac{3}{4} = \frac{21}{20}$

$\frac{7 \cdot 3}{5 \cdot 4} = \frac{21}{20}$

$x = \frac{7}{5}$

67. $x \div \frac{9}{5} = \frac{20}{63}$

$x \times \frac{5}{9} = \frac{20}{63}$

$\frac{4 \cdot 5}{7 \cdot 9} = \frac{20}{63}$

$x = \frac{4}{7}$

69. $20\frac{1}{4} \div 9 = \frac{81}{4} \div \frac{9}{1}$

$= \frac{81}{4} \times \frac{1}{9}$

$= \frac{9 \times 9 \times 1}{4 \times 9}$

$= \frac{9}{4}$

$= 2\frac{1}{4}$ gallons

71. $135 \div 3\frac{1}{3} = \frac{125}{1} \div \frac{10}{3}$

$= \frac{125}{1} \times \frac{3}{10}$

$= \frac{5 \times 25 \times 3}{5 \times 2}$

$= \frac{75}{2}$

$= 37\frac{1}{2}$ miles per hour

73. $38\frac{2}{3} \div \frac{2}{3} = \frac{116}{3} \div \frac{2}{3}$

$= \frac{116}{3} \times \frac{3}{2}$

$= \frac{58 \times 2 \times 3}{3 \times 2}$

$= 58$ students

75. $\frac{150}{1\frac{1}{2}} = \frac{150}{\frac{3}{2}} = \frac{150}{1} \times \frac{2}{3}$

$= 100$ large Styrofoam cups

77. $4\frac{3}{4} \div \frac{5}{6} = \frac{19}{4} \times \frac{6}{5} = \frac{19 \times 2 \times 3}{2 \times 2 \times 5}$

$= \frac{57}{10} = 5\frac{7}{10}$ attempts

It took six drill attempts.

79. $15 \div 5 = 3$

Exact $= 14\frac{2}{3} \div 5\frac{1}{6}$

$= \frac{44}{3} \times \frac{6}{31}$

$= \frac{88}{31} = 2\frac{26}{31}$

It is off by only $\frac{5}{31}$.

Cumulative Review

81. $39{,}576{,}304$ = Thirty-nine million, five hundred seventy-six thousand, three hundred four.

83. $126 + 34 + 9 + 891 + 12 + 27 = 1099$

How Am I Doing? Sections 2.1 - 2.5

1. $\frac{3}{8}$

2.

$$\begin{array}{r} 3500 \\ 2600 \\ +\ \ 800 \\ \hline 6900 \end{array}$$ Total

$\frac{800}{6900} = \frac{8}{69}$

3. $\frac{5}{124}$

4. $\frac{3}{18} = \frac{3 \div 3}{18 \div 3} = \frac{1}{6}$

5. $\frac{13}{39}=\frac{16\div 16}{112\div 16}=\frac{1}{7}$

6. $\frac{16}{112}=\frac{16\div 16}{112\div 16}=\frac{1}{7}$

7. $\frac{175}{200}=\frac{175\div 25}{200\div 25}=\frac{7}{8}$

8. $\frac{44}{121}=\frac{44\div 11}{121\div 11}=\frac{4}{11}$

9. $3\frac{2}{3}=\frac{3\times 3+2}{3}=\frac{11}{3}$

10. $6\frac{1}{9}=\frac{6\times 9+1}{9}=\frac{55}{9}$

11. $\begin{array}{r} 24 \\ 4\overline{)97} \\ \underline{8} \\ 17 \\ \underline{16} \\ 1 \end{array}$

$\frac{97}{4}=24\frac{1}{4}$

12. $\begin{array}{r} 5 \\ 5\overline{)29} \\ \underline{25} \\ 4 \end{array}$

$\frac{29}{5}=5\frac{4}{5}$

13. $\begin{array}{r} 2 \\ 17\overline{)36} \\ \underline{34} \\ 2 \end{array}$

$\frac{36}{17}=2\frac{2}{17}$

14. $\frac{5}{11}\times\frac{1}{4}=\frac{5}{44}$

15. $\frac{3}{7}\times\frac{14}{9}=\frac{3\times 2\times 7}{7\times 3\times 3}=\frac{2}{3}$

16. $12\frac{1}{3}\times 5\frac{1}{2}=\frac{37}{3}\times\frac{11}{2}$

$=\frac{407}{6}$

$=67\frac{5}{6}$

17. $\frac{3}{7}\div\frac{3}{7}=\frac{3}{7}\times\frac{7}{3}=1$

18. $\frac{7}{16}\div\frac{7}{8}=\frac{7}{16}\times\frac{8}{7}$

$=\frac{7\times 8}{2\times 8\times 7}$

$=\frac{1}{2}$

19. $6\frac{4}{7}\div 1\frac{5}{21}=\frac{46}{7}\div\frac{26}{21}$

$=\frac{46}{7}\times\frac{21}{26}$

$=\frac{2\times 23\times 3\times 7}{7\times 2\times 13}$

$=\frac{69}{13}$

$=5\frac{4}{13}$

20. $8\div\frac{12}{7}=\frac{8}{1}\times\frac{7}{12}$

$=\frac{2\times 4\times 7}{1\times 3\times 4}$

$=\frac{14}{3}$

$=4\frac{2}{3}$

Test on Sections 2.1-2.5

1. $\frac{23}{32}$

2. $\frac{112}{340}$

3. $\frac{19}{38}=\frac{19\div 19}{38\div 19}=\frac{1}{2}$

4. $\frac{35}{75}=\frac{35\div 5}{75\div 5}=\frac{7}{15}$

5. $\frac{24}{66}=\frac{24\div 6}{66\div 6}=\frac{4}{11}$

6. $\frac{125}{155}=\frac{125\div 5}{155\div 5}=\frac{25}{31}$

7. $\frac{39}{52}=\frac{39\div 13}{52\div 13}=\frac{3}{4}$

8. $\frac{84}{36}=\frac{84\div 12}{36\div 12}=\frac{7}{3}$ or $2\frac{1}{3}$

9. $3\frac{7}{12}=\frac{12\times 3+7}{12}=\frac{43}{12}$

10. $4\frac{1}{8}=\frac{8\times 4+1}{8}=\frac{33}{8}$

11. $7\overline{)45}$ quotient 6, $\underline{42}$, remainder 3; $\frac{45}{7}=6\frac{3}{7}$

12. $4\overline{)33}$ quotient 8, $\underline{32}$, remainder 1; $\frac{33}{4}=8\frac{1}{4}$

13. $\frac{3}{8}\times\frac{7}{11}=\frac{21}{88}$

14. $\frac{15}{7}\times\frac{3}{5}=\frac{15\times 3}{7\times 5}=\frac{3\times 5\times 3}{7\times 5}$
$=\frac{3\times 3}{7}=\frac{9}{7}=1\frac{2}{7}$

15. $18\times\frac{5}{6}=\frac{18}{1}\times\frac{5}{6}=\frac{3\times 6\times 5}{6}=3\times 5=15$

16. $\frac{3}{8}\times 44=\frac{3}{8}\times\frac{44}{1}=\frac{3\times 4\times 11}{2\times 4}$
$=\frac{3\times 11}{2}=\frac{33}{2}=16\frac{1}{2}$

17. $2\frac{1}{3}\times 5\frac{3}{4}=\frac{7}{3}\times\frac{23}{4}=\frac{161}{12}=13\frac{5}{12}$

18. $1\frac{3}{7}\times 3\frac{1}{3}=\frac{10}{7}\times\frac{10}{3}=\frac{10\times 10}{7\times 3}$
$=\frac{110}{21}=4\frac{16}{21}$

19. $\frac{4}{7}\div\frac{3}{4}=\frac{4}{7}\times\frac{4}{3}=\frac{4\times 4}{7\times 3}=\frac{16}{21}$

20. $\frac{8}{9}\div\frac{1}{6}=\frac{8}{9}\times\frac{6}{1}=\frac{8\times 3\times 2}{3\times 3}$
$=\frac{8\times 2}{3}=\frac{16}{3}=5\frac{1}{3}$

21. $5\frac{1}{4}\div\frac{3}{4}=\frac{21}{4}\times\frac{4}{3}=\frac{7\times 3\times 4}{4\times 3}=7$

22. $5\frac{3}{5}\div\frac{1}{2}=\frac{28}{5}\times\frac{1}{2}=\frac{28}{5}\times\frac{2}{1}$
$=\frac{28\times 2}{5}$
$=\frac{56}{5}$
$=11\frac{1}{5}$

23. $2\frac{1}{4}\times 3\frac{1}{2}=\frac{9}{4}\times\frac{7}{2}=\frac{63}{8}=7\frac{7}{8}$

24. $6\times 2\frac{1}{3}=\frac{6}{1}\times\frac{7}{3}=\frac{2\times 3\times 7}{3}=2\times 7=14$

25. $5 \div 1\frac{7}{8} = 5 \div \frac{15}{8} = \frac{5}{1} \times \frac{8}{15}$
$= \frac{5 \times 8}{5 \times 3} = \frac{8}{3}$
$= 2\frac{2}{3}$

26. $5\frac{3}{4} \div 2 = \frac{23}{4} \div 2 = \frac{23}{4} \times \frac{1}{2}$
$= \frac{23}{4 \times 2}$
$= \frac{23}{8}$
$= 2\frac{7}{8}$

27. $\frac{13}{20} \div \frac{4}{5} = \frac{13}{20} \times \frac{5}{4} = \frac{13 \times 5}{5 \times 4 \times 4} = \frac{13}{16}$

28. $\frac{4}{7} \div 8 = \frac{4}{7} \times \frac{1}{8} = \frac{4}{7 \times 2 \times 4} = \frac{1}{14}$

29. $\frac{9}{22} \times \frac{11}{16} = \frac{9 \times 11}{2 \times 11 \times 16} = \frac{9}{32}$

30. $\frac{14}{25} \times \frac{65}{42} = \frac{7 \times 2 \times 13 \times 5}{5 \times 5 \times 2 \times 3 \times 7} = \frac{13}{5 \times 3} = \frac{13}{15}$

31. $5\frac{1}{4} \times 8\frac{3}{4} = \frac{21}{4} \times \frac{35}{4}$
$= \frac{735}{16} = 45\frac{15}{16}$ square feet

32. $2\frac{2}{3} \times 1\frac{1}{2} = \frac{8}{3} \times \frac{3}{2} = \frac{4 \times 2 \times 3}{3 \times 2}$
$= 4$ cups

33. $62\frac{1}{2} \times \frac{3}{4} = \frac{125}{2} \times \frac{3}{4} = \frac{375}{8}$ or $46\frac{7}{8}$

She drove $46\frac{7}{8}$ miles

34. $12\frac{3}{8} \div \frac{3}{4} = \frac{99}{8} \times \frac{4}{3}$
$= \frac{3 \times 33 \times 4}{4 \times 2 \times 3}$
$= \frac{33}{2}$
$= 16\frac{1}{2}$ packages

He had 16 full packages with $\frac{3}{4} \times \frac{1}{2} = \frac{3}{8}$ pounds left over.

35. $136 \times \frac{3}{8} = \frac{136}{1} \times \frac{3}{8} = \frac{17 \times 8 \times 3}{8}$
$= 17 \times 3 = 51$ computers

36. $12,000 \div \frac{3}{5} = \frac{12,000}{1} \times \frac{5}{3}$
$= 4000 \times 5$
$= 20,000$ homes

$20,000 - 12,000 = 8000$ homes to be inspected.

37. $132 \div 8\frac{1}{4} = 132 \div \frac{33}{4} = \frac{132}{1} \times \frac{4}{33}$
$= \frac{33 \times 4 \times 4}{33} = 4 \times 4$
$= 16$ hours

38. $56\frac{1}{2} \div 8\frac{1}{4} = \frac{113}{2} \div \frac{33}{4} = \frac{113}{2} \times \frac{4}{33}$
$= \frac{113 \times 2 \times 2}{2 \times 33}$
$= \frac{226}{33} = 6\frac{28}{33}$

He can make 6 full tents, with

$8\frac{1}{4} \times \frac{28}{33} = \frac{33}{4} \times \frac{28}{33} = 7$ yards left over

39. $32\frac{4}{5} \div \frac{4}{5} = \frac{164}{5} \times \frac{5}{4} = 41$

He can use it for 41 days.

2.6 Exercises

1. $8 = 2\times2\times2$
$12 = 2\times2\times3$
$\text{LCM} = 2\times2\times2\times3 = 24$

3. $20 = 2\times2\times5$
$50 = 2\times5\times5$
$\text{LCM} = 2\times2\times5\times5 = 100$

5. $12 = 2\times2\times3$
$15 = 3\times5$
$\text{LCM} = 2\times2\times3\times5 = 60$

7. $9 = 3\times3$
$36 = 2\times2\times3\times3$
$\text{LCM} = 2\times2\times3\times3 = 36$

9. $21 = 3\times7$
$49 = 7\times7$
$\text{LCM} = 3\times7\times7 = 147$

11. $5 = 5$
$10 = 2\times5$
$\text{LCD} = 2\times5 = 10$

13. $7 = 7$
$4 = 2\times2$
$\text{LCD} = 2\times2\times7 = 28$

15. $5 = 5$
$7 = 7$
$\text{LCD} = 5\times7 = 35$

17. $9 = 3\times3$
$6 = 2\times3$
$\text{LCD} = 2\times3\times3 = 18$

19. $12 = 2\times2\times3$
$15 = 3\times5$
$\text{LCD} = 2\times2\times3\times5 = 60$

21. $4 = 2\times2$
$32 = 2\times2\times2\times2\times2$
$\text{LCD} = 2\times2\times2\times2\times2 = 32$

23. $10 = 2\times5$
$45 = 3\times3\times5$
$\text{LCD} = 2\times3\times3\times5 = 90$

25. $12 = 2\times2\times3$
$30 = 2\times3\times5$
$\text{LCD} = 2\times2\times3\times5 = 60$

27. $21 = 3\times7$
$35 = 5\times7$
$\text{LCD} = 3\times5\times7 = 105$

29. $18 = 2\times3\times3$
$45 = 3\times3\times5$
$\text{LCD} = 2\times3\times3\times5 = 90$

31. $3 = 3$
$2 = 2$
$6 = 2\times3$
$\text{LCD} = 2\times3 = 6$

33. $6 = 2\times3$
$10 = 2\times5$
$4 = 2\times2$
$\text{LCD} = 2\times2\times3\times5 = 60$

35. $11 = 11$
$12 = 2\times2\times3$
$6 = 2\times3$
$\text{LCD} = 2\times2\times3\times11 = 132$

37. $12 = 2\times2\times3$
$21 = 3\times7$
$14 = 2\times7$
$\text{LCD} = 2\times2\times3\times7 = 84$

39. $15 = 3\times5$
$12 = 2\times2\times3$
$8 = 2\times2\times2$
$\text{LCD} = 2\times2\times2\times3\times5 = 120$

41. $\frac{1}{3} = \frac{1}{3} \times \frac{3}{3} = \frac{3}{9}$

3

43. $\frac{5}{7} = \frac{5 \times 7}{7 \times 7} = \frac{35}{49}$

35

45. $\frac{4}{11} = \frac{4}{11} \times \frac{5}{5} = \frac{20}{55}$

20

47. $\frac{7}{24} = \frac{7}{24} \times \frac{2}{2} = \frac{14}{48}$

14

49. $\frac{8}{9} = \frac{8}{9} \times \frac{12}{12} = \frac{96}{108}$

96

51. $\frac{7}{20} = \frac{7}{20} \times \frac{9}{9} = \frac{63}{180}$

63

53. $\frac{7}{12} = \frac{7 \times 3}{12 \times 3} = \frac{21}{36}$

$\frac{5}{9} = \frac{5 \times 4}{9 \times 4} = \frac{20}{36}$

55. $\frac{5}{16} \times \frac{5}{5} = \frac{25}{80}$

$\frac{17}{20} \times \frac{4}{4} = \frac{68}{80}$

57. $\frac{9}{10} = \frac{9 \times 2}{10 \times 2} = \frac{18}{20}$

$\frac{19}{20} = \frac{19}{20}$

59. $5 = 5$

$35 = 5 \times 7$

$\text{LCD} = 5 \times 7 = 35$

$\frac{2}{5} = \frac{2 \times 7}{5 \times 7} = \frac{14}{35}$

$\frac{14}{35}$ and $\frac{9}{35}$

61. $24 = 2^3 \times 3$

$8 = 2^3$

$\text{LCD} = 2^3 \times 3 = 24$

$\frac{5}{24}$

$\frac{3}{8} = \frac{3 \times 3}{8 \times 3} = \frac{9}{24}$

$\frac{9}{24}$ and $\frac{5}{24}$

63. $6 = 2 \times 3$

$10 = 2 \times 5$

$\text{LCD} = 2 \times 3 \times 5 = 30$

$\frac{7}{10} \times \frac{3}{3} = \frac{21}{30}$ $\quad \frac{5}{6} \times \frac{5}{5} = \frac{25}{30}$

65. $15 = 3 \times 5$

$12 = 2^2 \times 3$

$\text{LCD} = 3 \times 5 \times 2^2 = 60$

$\frac{4}{15} = \frac{4 \times 4}{15 \times 4} = \frac{16}{60}$

$\frac{5}{12} = \frac{5 \times 5}{12 \times 5} = \frac{25}{60}$

$\frac{16}{60}$ and $\frac{25}{60}$

67. $18 = 2 \times 3 \times 3$

$36 = 2 \times 2 \times 3 \times 3$

$12 = 2 \times 2 \times 3$

$\text{LCD} = 2 \times 2 \times 3 \times 3 = 36$

$\frac{5}{18} \times \frac{2}{2} = \frac{10}{36}$ $\quad \frac{11}{36} \times \frac{11}{36}$ $\quad \frac{7}{12} \times \frac{3}{3} = \frac{21}{36}$

69. $56 = 2\times2\times2\times7$

$8 = 2\times2\times2$

$7 = 7$

$\text{LCD} = 2\times2\times2\times7 = 56$

$\frac{3}{56}, \frac{7}{8}\times\frac{7}{7}, \frac{5}{7}\times\frac{8}{8}$

$\frac{3}{56}, \frac{49}{56}, \frac{40}{56}$

71. $63 = 3\times3\times7$

$21 = 3\times7$

$9 = 3\times3$

$\text{LCD} = 3\times3\times7 = 63$

$\frac{5}{63}, \frac{4}{21}\times\frac{3}{3}, \frac{8}{9}\times\frac{7}{7}$

$\frac{5}{63}, \frac{12}{63}, \frac{56}{63}$

73. a. $16 = 2\times2\times2\times2$

$4 = 2\times2$

$8 = 2\times2\times2$

$\text{LCD} = 2\times2\times2\times2 = 16$

b. $\frac{3}{16}, \frac{3}{4}\times\frac{4}{4}, \frac{3}{8}\times\frac{2}{2}$

$\frac{3}{16}, \frac{12}{16}, \frac{6}{16}$

Cumulative Review

75.

$$\begin{array}{r} 178 \\ 32\overline{)5699} \\ \underline{32} \\ 249 \\ \underline{224} \\ 259 \\ \underline{256} \\ 3 \end{array}$$

178 R 3

77.

$$\begin{aligned}(5-3)^2+4\times6-3 &= 2^2+4\times6-3\\ &= 4+4\times6-3\\ &= 4+24-3\\ &= 25\end{aligned}$$

79.

$$\begin{array}{rrrr} 11,205 & 13,359 & 15,569 & 17,603 \\ -\ 8,414 & -\ 11,205 & -\ 13,359 & -\ 15,569 \\ \hline 2,791 & 2,154 & 2,210 & 2,034 \end{array}$$

Smallest increase is $2,034 between 1995 and 2000.

81.

$$\begin{array}{r} 17,603 \\ -\ \ 9700 \\ \hline 7903 \end{array}$$

He would have to earn more than $7903 per year.

83.

$$\begin{array}{r} 17,603 \\ -\ 13,359 \\ \hline 4,244 \end{array}$$

For year 2010, the poverty level is

$$\begin{array}{r} 17,603 \\ +\ 4,244 \\ \hline \$21,847 \end{array}$$

2.7 Exercises

1. $\frac{5}{9}+\frac{2}{9}=\frac{5+2}{9}=\frac{7}{9}$

3. $\frac{7}{18}+\frac{15}{18}=\frac{22}{18}=\frac{11}{9}$ or $1\frac{2}{9}$

5. $\frac{5}{24}-\frac{3}{24}=\frac{2}{24}=\frac{1}{12}$

7. $\frac{53}{88}-\frac{19}{88}=\frac{34}{88}=\frac{17}{44}$

9.

$$\begin{aligned}\frac{1}{3}+\frac{1}{2} &= \frac{1}{3}\times\frac{2}{2}+\frac{1}{2}\times\frac{3}{3}\\ &= \frac{2}{6}+\frac{3}{6}\\ &= \frac{5}{6}\end{aligned}$$

11. $\frac{3}{10}+\frac{3}{20}=\frac{3}{10}\times\frac{2}{2}+\frac{3}{20}$
$=\frac{6}{20}+\frac{3}{20}$
$=\frac{9}{20}$

13. $\frac{1}{8}+\frac{3}{4}=\frac{1}{8}+\frac{3}{4}\times\frac{2}{2}$
$=\frac{1}{8}+\frac{6}{6}$
$=\frac{7}{8}$

15. $\frac{4}{5}+\frac{7}{20}=\frac{4}{5}\times\frac{4}{4}+\frac{7}{20}$
$=\frac{16}{20}+\frac{7}{20}=\frac{23}{20}=1\frac{3}{20}$
$=1\frac{3}{20}$

17. $\frac{3}{10}+\frac{7}{100}=\frac{3}{10}\times\frac{10}{10}+\frac{7}{100}$
$=\frac{30}{100}+\frac{7}{100}$
$=\frac{37}{100}$

19. $\frac{3}{25}+\frac{1}{35}=\frac{3}{25}\times\frac{7}{7}+\frac{1}{35}\times\frac{5}{5}$
$=\frac{21}{175}+\frac{5}{175}$
$=\frac{26}{175}$

21. $\frac{7}{8}+\frac{5}{12}=\frac{7}{8}\times\frac{3}{3}+\frac{5}{12}\times\frac{2}{2}$
$=\frac{21}{24}+\frac{10}{24}=\frac{31}{24}$
$=1\frac{7}{24}$

23. $\frac{3}{8}+\frac{3}{10}=\frac{3}{8}\times\frac{5}{5}+\frac{3}{10}\times\frac{4}{4}$
$=\frac{15}{40}+\frac{12}{40}$
$=\frac{27}{40}$

25. $\frac{5}{12}-\frac{1}{6}=\frac{5}{12}-\frac{1}{6}\times\frac{2}{2}$
$=\frac{5}{12}-\frac{2}{12}=\frac{3}{12}$
$=\frac{1}{4}$

27. $\frac{3}{7}-\frac{1}{5}=\frac{3}{7}\times\frac{5}{5}-\frac{1}{5}\times\frac{7}{7}$
$=\frac{15}{35}-\frac{7}{35}$
$=\frac{8}{35}$

29. $\frac{5}{9}-\frac{5}{36}=\frac{5}{9}\times\frac{4}{4}-\frac{5}{36}$
$=\frac{20}{36}-\frac{5}{36}$
$=\frac{15}{36}=\frac{5}{12}$

31. $\frac{5}{12}-\frac{7}{30}=\frac{5}{12}\times\frac{5}{5}-\frac{7}{30}\times\frac{2}{2}$
$=\frac{25}{60}-\frac{14}{60}$
$=\frac{11}{60}$

33. $\frac{11}{12}-\frac{2}{3}=\frac{11}{12}-\frac{2}{3}\times\frac{4}{4}$
$=\frac{11}{12}-\frac{8}{12}$
$=\frac{3}{12}=\frac{1}{4}$

35. $\frac{17}{21}-\frac{1}{7}=\frac{17}{21}-\frac{1}{7}\times\frac{3}{3}$
$=\frac{17}{21}-\frac{3}{21}$
$=\frac{14}{21}$
$=\frac{2}{3}$

37. $\frac{7}{24}-\frac{1}{6}=\frac{7}{24}-\frac{1}{6}\times\frac{4}{4}$
$=\frac{7}{24}-\frac{4}{24}$
$=\frac{3}{24}$
$=\frac{1}{8}$

39. $\frac{10}{16}-\frac{5}{8}=\frac{10}{16}-\frac{5}{8}\times\frac{2}{2}$
$=\frac{10}{16}-\frac{10}{16}=0$

41. $\frac{23}{36}-\frac{2}{9}=\frac{23}{36}-\frac{2}{9}\times\frac{4}{4}$
$=\frac{23}{36}-\frac{8}{36}=\frac{15}{36}$
$=\frac{5}{12}$

43. $\frac{1}{2}+\frac{2}{7}+\frac{3}{14}=\frac{1}{2}\times\frac{7}{7}+\frac{2}{7}\times\frac{2}{2}+\frac{3}{14}$
$=\frac{7}{14}+\frac{4}{14}+\frac{3}{14}$
$=\frac{14}{14}$
$=1$

45. $\frac{5}{30}+\frac{3}{40}+\frac{1}{8}=\frac{5}{30}\times\frac{4}{4}+\frac{3}{40}\times\frac{3}{3}+\frac{1}{8}\times\frac{15}{15}$
$=\frac{20}{120}+\frac{9}{120}+\frac{15}{120}$
$=\frac{44}{120}$
$=\frac{11}{30}$

47. $\frac{7}{30}+\frac{2}{5}+\frac{5}{6}=\frac{7}{30}+\frac{2}{5}\times\frac{6}{6}+\frac{5}{6}\times\frac{5}{5}$
$=\frac{7}{30}+\frac{12}{30}+\frac{25}{30}$
$=\frac{44}{30}=\frac{22}{15}$
$=1\frac{7}{15}$

49. $x+\frac{1}{7}=\frac{5}{14}$
$x+\frac{1}{7}\times\frac{2}{2}=\frac{5}{14}$
$x+\frac{2}{14}=\frac{5}{14}$
$\frac{3}{14}+\frac{2}{14}=\frac{5}{14}$
$x=\frac{3}{14}$

51. $x+\frac{2}{3}=\frac{9}{11}$
$x+\frac{2}{3}\times\frac{11}{11}=\frac{9}{11}\times\frac{3}{3}$
$x+\frac{22}{33}=\frac{27}{33}$
$\frac{5}{33}+\frac{22}{33}=\frac{27}{33}$
$x=\frac{5}{33}$

53. $x - \frac{1}{5} = \frac{4}{12}$

$x - \frac{1}{5} \times \frac{12}{12} = \frac{4}{12} \times \frac{5}{5}$

$x - \frac{12}{60} = \frac{20}{60}$

$\frac{32}{60} - \frac{12}{60} = \frac{20}{60}$

$x = \frac{32}{60} = \frac{8}{15}$

55. $\frac{3}{4} + \frac{2}{3} = \frac{3}{4} \times \frac{3}{3} + \frac{2}{3} \times \frac{4}{4}$

$= \frac{9}{12} + \frac{8}{12} = \frac{17}{12}$

$= 1\frac{5}{12}$ cups

57. $\frac{2}{3} + \frac{5}{6} = \frac{2}{3} \times \frac{2}{2} + \frac{5}{6}$

$= \frac{4}{6} + \frac{5}{6}$

$= \frac{9}{6} = \frac{3}{2} = 1\frac{1}{2}$ pounds

59. $\frac{11}{12} - \frac{3}{5} = \frac{11}{12} \times \frac{5}{5} - \frac{3}{5} \times \frac{12}{12}$

$= \frac{55}{60} - \frac{36}{60}$

$= \frac{19}{60}$ of the book report

61. Before he ate half, there were

$6 \div \frac{1}{2} = 6 \times 2 = 12$

chocolates. While walking, he ate $\frac{1}{4}$ of the chocolates, leaving

$1 - \frac{1}{4} = \frac{3}{4}$

in the box. The box had

$12 \div \frac{3}{4} = \frac{12}{1} \times \frac{4}{3} = 16$ chocolates

63. $\frac{5}{6} - \frac{9}{14} = \frac{35}{42} - \frac{27}{42}$

$= \frac{8}{42}$

$= \frac{4}{21}$ of the membership

Cumulative Review

65. $\frac{15}{85} = \frac{15 \div 5}{85 \div 5} = \frac{3}{17}$

67. $14\overline{)125}$ quotient 8

$\underline{112}$

13

$\frac{125}{14} = 8\frac{13}{14}$

69. $45\frac{1}{2} \times 4\frac{3}{4} = \frac{11}{2} \times \frac{19}{4} = \frac{209}{8} = 26\frac{1}{8}$

2.8 Exercises

1. $7\frac{1}{8}$

$+2\frac{5}{8}$

$9\frac{6}{8} = 9\frac{3}{4}$

3. $15\frac{3}{14}$

$-11\frac{1}{14}$

$4\frac{2}{14} = 4\frac{1}{7}$

5. $12\frac{1}{3}$ $+ 5\frac{1}{6}$ $12\frac{2}{6}$ $+ 5\frac{1}{6}$

$17\frac{3}{6} = 17\frac{1}{2}$

7. $5\frac{4}{5}$ $+ 10\frac{3}{10}$ $5\frac{8}{10}$ $+ 10\frac{3}{10}$

$15\frac{11}{10} = 16\frac{1}{10}$

9. $\begin{array}{r} 1 \\ -\frac{3}{7} \\ \hline \end{array}$ $\begin{array}{r} \frac{7}{7} \\ -\frac{3}{7} \\ \hline \frac{4}{7} \end{array}$

11. $\begin{array}{r} 1\frac{5}{6} \\ +\frac{7}{8} \\ \hline \end{array}$ $\begin{array}{r} 1\frac{20}{24} \\ +\frac{21}{24} \\ \hline 1\frac{41}{24} = 2\frac{17}{24} \end{array}$

13. $\begin{array}{r} 6\frac{1}{5} \\ +3\frac{1}{6} \\ \hline \end{array}$ $\begin{array}{r} 6\frac{6}{30} \\ +3\frac{5}{30} \\ \hline 9\frac{11}{30} \end{array}$

15. $\begin{array}{r} 8\frac{1}{4} \\ -8\frac{4}{16} \\ \hline \end{array}$ $\begin{array}{r} 8\frac{4}{16} \\ -8\frac{4}{16} \\ \hline 0 \end{array}$

17. $\begin{array}{r} 12\frac{1}{3} \\ -7\frac{2}{5} \\ \hline \end{array}$ $\begin{array}{r} 12\frac{5}{15} \\ -7\frac{6}{15} \\ \hline \end{array}$ $\begin{array}{r} 11\frac{20}{15} \\ -7\frac{6}{15} \\ \hline 4\frac{14}{15} \end{array}$

19. $\begin{array}{r} 30 \\ -15\frac{3}{7} \\ \hline \end{array}$ $\begin{array}{r} 29\frac{7}{7} \\ -15\frac{3}{7} \\ \hline 14\frac{4}{7} \end{array}$

21. $\begin{array}{r} 3 \\ +4\frac{2}{5} \\ \hline 7\frac{2}{5} \end{array}$

23. $\begin{array}{r} 45 \\ -35\frac{2}{5} \\ \hline \end{array}$ $\begin{array}{r} 44\frac{5}{5} \\ -35\frac{2}{5} \\ \hline 9\frac{3}{5} \end{array}$

25. $\begin{array}{r} 15\frac{4}{15} \\ +26\frac{8}{15} \\ \hline 41\frac{12}{15} = 41\frac{4}{5} \end{array}$

27. $\begin{array}{r} 4\frac{1}{3} \\ +2\frac{1}{4} \\ \hline \end{array}$ $\begin{array}{r} 4\frac{4}{12} \\ +2\frac{3}{12} \\ \hline 6\frac{7}{12} \end{array}$

29. $\begin{array}{r} 3\frac{3}{4} \\ +4\frac{5}{12} \\ \hline \end{array}$ $\begin{array}{r} 3\frac{9}{12} \\ +4\frac{5}{12} \\ \hline 7\frac{14}{12} = 8\frac{2}{12} = 8\frac{1}{6} \end{array}$

31. $\begin{array}{r} 47\frac{3}{10} \\ +26\frac{5}{8} \\ \hline \end{array}$ $\begin{array}{r} 47\frac{12}{40} \\ +26\frac{25}{40} \\ \hline 73\frac{37}{40} \end{array}$

33. $\begin{array}{r} 19\frac{5}{6} \\ -14\frac{1}{3} \\ \hline \end{array}$ $\begin{array}{r} 19\frac{5}{6} \\ -14\frac{2}{6} \\ \hline 5\frac{3}{6} = 5\frac{1}{2} \end{array}$

35. $\begin{array}{r} 6\frac{1}{12} \\ -5\frac{10}{24} \\ \hline \end{array}$ $\begin{array}{r} 6\frac{2}{24} \\ -5\frac{10}{24} \\ \hline \end{array}$ $\begin{array}{r} 5\frac{26}{36} \\ -5\frac{10}{24} \\ \hline \frac{16}{24} = \frac{2}{3} \end{array}$

37. $\begin{array}{r} 12\frac{3}{20} \\ -7\frac{7}{15} \\ \hline \end{array}$ $\begin{array}{r} 12\frac{9}{60} \\ -7\frac{28}{60} \\ \hline \end{array}$ $\begin{array}{r} 11\frac{69}{60} \\ -7\frac{28}{60} \\ \hline 4\frac{41}{60} \end{array}$

39. $\begin{array}{r} 12 \\ -3\frac{7}{15} \\ \hline \end{array}$ $\begin{array}{r} 11\frac{15}{15} \\ -3\frac{7}{15} \\ \hline 8\frac{8}{15} \end{array}$

41. $\begin{array}{r} 120 \\ -17\frac{3}{8} \\ \hline \end{array}$ $\begin{array}{r} 119\frac{8}{8} \\ -17\frac{3}{8} \\ \hline 102\frac{5}{8} \end{array}$

43. $\begin{array}{r} 3\frac{5}{8} \\ 2\frac{2}{3} \\ +7\frac{3}{4} \\ \hline \end{array}$ $\begin{array}{r} 3\frac{15}{24} \\ 2\frac{16}{24} \\ +7\frac{18}{24} \\ \hline 12\frac{49}{24} = 14\frac{1}{24} \end{array}$

45. $\begin{array}{r} 20\tfrac{3}{4} \\ +\ 22\tfrac{3}{8} \\ \hline \end{array}$ $\begin{array}{r} 20\tfrac{6}{8} \\ +22\tfrac{3}{8} \\ \hline 42\tfrac{9}{8} = 43\tfrac{1}{8} \end{array}$

$43\frac{1}{8}$ miles

47. $\begin{array}{r} 6\tfrac{3}{8} \\ -\ 4\tfrac{1}{3} \\ \hline \end{array}$ $\begin{array}{r} 6\tfrac{9}{24} \\ -\ 4\tfrac{8}{24} \\ \hline 2\tfrac{1}{24} \end{array}$

$2\frac{1}{24}$ pounds

49. $\begin{array}{r} 72\tfrac{1}{2} \\ -\ 69\tfrac{3}{4} \\ \hline \end{array}$ $\begin{array}{r} 72\tfrac{2}{4} \\ -\ 69\tfrac{3}{4} \\ \hline \end{array}$ $\begin{array}{r} 71\tfrac{6}{4} \\ -\ 69\tfrac{3}{4} \\ \hline 2\tfrac{3}{4} \end{array}$

$2\frac{3}{4}$ inches

51. a. $\begin{array}{r} 1\tfrac{3}{4} \\ +\ 2\tfrac{1}{6} \\ \hline \end{array}$ $\begin{array}{r} 1\tfrac{9}{12} \\ +2\tfrac{2}{12} \\ \hline 3\tfrac{11}{12} \end{array}$ pounds

51. b. $\begin{array}{r} 8 \\ -\ 3\tfrac{11}{12} \\ \hline \end{array}$ $\begin{array}{r} 7\tfrac{12}{12} \\ -3\tfrac{11}{12} \\ \hline 4\tfrac{1}{12} \end{array}$ pounds

53. $\frac{379}{8}+\frac{89}{5}=\frac{1895}{40}+\frac{712}{40}$
$=\frac{2607}{40}$
$=65\frac{7}{40}$

55. Estimate: $35+24=59$

Exact: $\begin{array}{r} 35\tfrac{1}{6} \\ +\ 24\tfrac{25}{12} \\ \hline \end{array}$ $\begin{array}{r} 35\tfrac{2}{12} \\ +\ 24\tfrac{5}{12} \\ \hline 59\tfrac{7}{12} \end{array}$

57. $\frac{6}{7}-\frac{4}{7}\times\frac{1}{3}=\frac{6}{7}-\frac{4}{21}$
$=\frac{6}{7}\times\frac{3}{3}-\frac{4}{21}$
$=\frac{18}{21}-\frac{4}{21}=\frac{14}{21}$
$=\frac{2}{3}$

59. $\frac{1}{2}+\frac{3}{8}\div\frac{3}{4}=\frac{1}{2}+\frac{3}{8}\times\frac{4}{3}$
$=\frac{1}{2}+\frac{1}{2}=\frac{2}{2}$
$=1$

61. $\frac{5}{7}\times\frac{7}{2}\div\frac{3}{2}=\frac{5}{7}\times\frac{7}{2}\times\frac{2}{3}=\frac{5}{3}=1\frac{2}{3}$

63. $\frac{3}{5}\times\frac{1}{2}+\frac{1}{5}\div\frac{2}{3}=\frac{3}{5}\times\frac{1}{2}+\frac{1}{5}\times\frac{3}{2}$
$=\frac{3}{10}+\frac{3}{10}=\frac{6}{10}$
$=\frac{3}{5}$

65. $\left(\frac{3}{5}-\frac{3}{20}\right)\times\frac{4}{5}=\left(\frac{12}{20}-\frac{3}{20}\right)\times\frac{4}{5}$
$=\frac{9}{20}\times\frac{4}{5}$
$=\frac{9}{25}$

67. $\left(\frac{4}{3}\right)^2\div\frac{11}{9}=\frac{16}{9}\times\frac{9}{11}=\frac{16}{11}$ or $1\frac{5}{11}$

69. $\frac{1}{4}\times\left(\frac{2}{3}\right)^2=\frac{1}{4}\times\frac{4}{9}=\frac{1}{9}$

71. $\frac{5}{6}\div\left(\frac{2}{3}+\frac{1}{6}\right)^2=\frac{5}{6}\div\left(\frac{4}{6}+\frac{1}{6}\right)^2$
$=\frac{5}{6}\div\left(\frac{5}{6}\right)^2$
$=\frac{5}{6}\div\frac{25}{36}$
$=\frac{5}{6}\times\frac{36}{25}$
$=\frac{6}{5}$ or $1\frac{1}{5}$

Cumulative Review

73. $\begin{array}{r} 1200 \\ \times\ 400 \\ \hline 480{,}000 \end{array}$

75. Changes are:
$25\times 27+520+30\times 8+2972$
$=675+520+240+2972$
$=\$4407$
Amount left over for appliances:
$6300-4407=\$1893$

2.9 Exercises

1. $\begin{array}{r} 5\frac{1}{4} \\ 2\frac{5}{6} \\ +\ 8\frac{7}{12} \\ \hline \end{array}$ $\quad$ $\begin{array}{r} 5\frac{3}{12} \\ 2\frac{10}{12} \\ +\ 8\frac{7}{12} \\ \hline 15\frac{20}{12}=16\frac{8}{12}=16\frac{2}{3} \text{ feet} \end{array}$

3. $\begin{array}{r} 7\frac{5}{6} \\ 8\frac{1}{8} \\ +\ 9\frac{1}{2} \\ \hline \end{array}$ $\quad$ $\begin{array}{r} 7\frac{20}{24} \\ 8\frac{3}{24} \\ +\ 9\frac{12}{24} \\ \hline 24\frac{35}{24}=25\frac{11}{24} \text{ tons} \end{array}$

5. $\frac{1}{16}+\frac{3}{4}+\frac{1}{16}+\frac{3}{16}+\frac{1}{2}$
$=\frac{1}{16}+\frac{12}{16}+\frac{1}{16}+\frac{3}{16}+\frac{8}{16}$
$=\frac{25}{16}$
$=1\frac{9}{16}$ inches

7. Sum:
$\begin{array}{r} 6\frac{3}{4} \\ +\ 9\frac{1}{2} \\ \hline \end{array}$ $\quad$ $\begin{array}{r} 6\frac{3}{4} \\ +\ 9\frac{2}{4} \\ \hline 15\frac{5}{4}=16\frac{1}{4} \end{array}$

Miles to end:
$\begin{array}{r} 26\frac{1}{5} \\ -\ 16\frac{1}{4} \\ \hline \end{array}$ $\quad$ $\begin{array}{r} 26\frac{4}{20} \\ -16\frac{5}{20} \\ \hline \end{array}$ $\quad$ $\begin{array}{r} 25\frac{24}{20} \\ -16\frac{5}{20} \\ \hline 9\frac{19}{20} \text{ miles} \end{array}$

9. $10\frac{1}{2}\times 8=\frac{21}{2}\times 8=84$
$10\frac{1}{2}\times 1\frac{1}{2}\times 4=\frac{21}{2}\times\frac{3}{2}\times 4=63$
Total $=84+63=\$147$

11. $36\frac{3}{4}\times 7\frac{1}{2}=\frac{147}{4}\times\frac{15}{2}$
$=\frac{2205}{8}$
$=275\frac{5}{8}$ gallons

13. $22\frac{1}{2}\times 4\frac{3}{4}=\frac{45}{2}\times\frac{19}{4}$
$=\frac{855}{8}$
$=106\frac{7}{8}$ nautical miles

15. $\frac{1}{5}+\frac{1}{15}+\frac{1}{20}=\frac{12}{60}+\frac{4}{60}+\frac{3}{60}$

$=\frac{12+4+3}{60}$

$=\frac{19}{60}$

$\frac{19}{60}(660)-209$, so \$209 is deducted.

$660-209=451$

She has \$451 per week left.

17. a. $13\frac{1}{2}\div\frac{2}{3}=\frac{27}{2}\times\frac{3}{2}=20.25$

20 bandanas

b. $\frac{1}{4}\times\frac{2}{3}=\frac{1}{6}$ ft left over

c. $2\frac{1}{2}\times 20=\frac{5}{2}\times 20=50$

\$50

19. a. $18\frac{1}{2}-1\frac{1}{4}-3\frac{1}{8}$

$=\frac{37}{2}-\frac{5}{4}-\frac{28}{8}=\frac{148}{8}-\frac{10}{8}-\frac{25}{8}$

$=\frac{148-10-25}{8}$

$=\frac{113}{8}$

$=14\frac{1}{8}$ ounces of bread

b.

$$\begin{array}{r} 14\frac{3}{4} \\ -\ 14\frac{1}{8} \\ \hline \end{array} \qquad \begin{array}{r} 14\frac{6}{8} \\ -\ 14\frac{1}{8} \\ \hline \frac{5}{8} \end{array}$$

$\frac{5}{8}$ of an ounce

21. a. $160\frac{1}{8}\div 5\frac{1}{4}=\frac{1281}{8}\div\frac{21}{4}$

$=\frac{1281}{8}\times\frac{4}{21}$

$=\frac{61}{2}$

$=30\frac{1}{2}$ knots

b. $213\frac{1}{2}\div\frac{61}{2}=\frac{427}{2}\div\frac{61}{2}$

$=\frac{427}{2}\times\frac{2}{61}$

$=\frac{427}{61}$

$=7$ hours

23. a. $6856\frac{1}{4}\div 1\frac{1}{4}=\frac{27,425}{4}\div\frac{5}{4}$

$=\frac{27,425}{4}\times\frac{4}{5}$

$=5485$ bushels

b. $6856\frac{1}{4}\times 1\frac{3}{4}=\frac{27,425}{4}\times\frac{7}{4}$

$=\frac{191,425}{16}$

$=11,998\frac{7}{16}$ cubic feet

c. $11,998\frac{7}{16}\div 1\frac{1}{4}=\frac{191,975}{16}\times\frac{4}{5}$

$=\frac{38,395}{4}$

$=9598\frac{3}{4}$ bushels

25. Salad this year:

$$\begin{array}{r} 12\frac{3}{4} \\ 10\frac{1}{3} \\ +\ 6\frac{8}{9} \\ \hline \end{array} \qquad \begin{array}{r} 12\frac{27}{36} \\ 10\frac{12}{36} \\ +\ 6\frac{32}{36} \\ \hline 28\frac{71}{36}=29\frac{35}{36} \end{array}$$

Salad last year:

$29\frac{35}{36}\times\frac{1}{2}=\frac{1079}{36}\times\frac{1}{2}=\frac{1079}{72}$

$=14\frac{71}{72}$ lb

Cumulative Review

27. $\begin{array}{r} 16{,}846 \\ 19{,}321 \\ +\ 8{,}078 \\ \hline 44{,}245 \end{array}$

29. $\begin{array}{r} 1683 \\ \times\ 27 \\ \hline 11781 \\ 3366 \\ \hline 45{,}441 \end{array}$

Putting Your Skills to Work

1. $\begin{array}{r} 2\frac{1}{8} \\ +\ 1\frac{1}{8} \\ \hline \end{array}$

$3\frac{2}{8} = 3\frac{1}{4}$ miles

2. $\begin{array}{r} 2\frac{1}{4} \\ +\ 3\frac{1}{8} \\ \hline \end{array}$ $\begin{array}{r} 2\frac{4}{8} \\ +\ 3\frac{1}{8} \\ \hline \end{array}$

$5\frac{3}{8}$ miles

3. $\begin{array}{r} 2\frac{5}{8} \\ 2\frac{1}{4} \\ +\ 2\frac{7}{8} \\ \hline \end{array}$ $\begin{array}{r} 2\frac{5}{8} \\ 2\frac{2}{8} \\ +\ 2\frac{7}{8} \\ \hline \end{array}$

$6\frac{14}{8} = 7\frac{3}{4}$ miles

4. $\begin{array}{r} 2\frac{5}{8} \\ 2\frac{1}{4} \\ 3\frac{1}{8} \\ +\ 1\frac{1}{4} \\ \hline \end{array}$ $\begin{array}{r} 2\frac{5}{8} \\ 2\frac{2}{8} \\ 3\frac{1}{8} \\ +\ 1\frac{2}{8} \\ \hline \end{array}$

$8\frac{10}{8} = 9\frac{1}{4}$ miles

$\begin{array}{r} 9\frac{1}{4} \\ -\ 7\frac{3}{4} \\ \hline \end{array}$ $\begin{array}{r} 8\frac{5}{4} \\ -\ 7\frac{3}{4} \\ \hline \end{array}$

$1\frac{2}{4} = 1\frac{1}{2}$ miles extra

5. $\begin{array}{r} 1\frac{3}{8} \\ +\ 2\frac{1}{8} \\ \hline 3\frac{4}{8} = 3\frac{1}{2} \end{array}$

$3\frac{1}{2} \div 8 = \frac{7}{2} \times \frac{1}{8}$

$= \frac{7}{16}$ hour

6. $\begin{array}{r} 3\frac{1}{8} \\ 2\frac{1}{4} \\ 3\frac{3}{8} \\ 2\frac{3}{4} \\ 1\frac{1}{2} \\ +\ 1\frac{1}{8} \\ \hline \end{array}$ $\begin{array}{r} 3\frac{1}{8} \\ 2\frac{2}{8} \\ 3\frac{3}{8} \\ 2\frac{6}{8} \\ 1\frac{4}{8} \\ +\ 1\frac{1}{8} \\ \hline 12\frac{17}{8} = 14\frac{1}{8} \end{array}$

$14\frac{1}{8} \div 6 = \frac{113}{8} \times \frac{1}{6} = \frac{113}{48} = 2\frac{17}{48}$ hours

$\begin{array}{r} 3 \\ -\ 2\frac{17}{48} \\ \hline \end{array}$ $\begin{array}{r} 2\frac{48}{48} \\ -2\frac{17}{48} \\ \hline \frac{31}{48} \end{array}$ hours left

$\frac{31}{48} \times 60 = \frac{155}{4} = 38\frac{3}{4}$

She will have $38\frac{3}{4}$ minutes left.

Chapter 2 Review Problems

1. $\frac{3}{8}$

2. $\frac{5}{12}$

3. $\frac{4}{7}$

4. $\frac{7}{9}$

5. $\frac{6}{31}$

6. $\frac{87}{100}$

7. $$\begin{aligned} 54 &= 2\times 27 \\ &= 2\times 3\times 9 \\ &= 2\times 3\times 3\times 3 \\ &= 2\times 3^3 \end{aligned}$$

8. $$\begin{aligned} 120 &= 10\times 12 \\ &= 2\times 5\times 2\times 2\times 3 \\ &= 2^3\times 3\times 5 \end{aligned}$$

9. $$\begin{aligned} 168 &= 8\times 21 \\ &= 2\times 2\times 2\times 3\times 7 \\ &= 2^3\times 3\times 7 \end{aligned}$$

10. Prime

11. $$\begin{aligned} 78 &= 2\times 39 \\ &= 2\times 3\times 13 \end{aligned}$$

12. Prime

13. $$\begin{aligned} \frac{12}{42} &= \frac{12\div 6}{42\div 6} \\ &= \frac{2}{7} \end{aligned}$$

14. $$\begin{aligned} \frac{13}{52} &= \frac{13\div 13}{52\div 13} \\ &= \frac{1}{4} \end{aligned}$$

15. $$\begin{aligned} \frac{21}{36} &= \frac{21\div 3}{36\div 3} \\ &= \frac{7}{12} \end{aligned}$$

16. $$\begin{aligned} \frac{26}{34} &= \frac{26\div 2}{34\div 2} \\ &= \frac{13}{17} \end{aligned}$$

17. $$\begin{aligned} \frac{168}{192} &= \frac{168\div 24}{192\div 24} \\ &= \frac{7}{8} \end{aligned}$$

18. $$\begin{aligned} \frac{51}{105} &= \frac{51\div 3}{105\div 3} \\ &= \frac{17}{35} \end{aligned}$$

19. $$\begin{aligned} 4\frac{3}{8} &= \frac{8\times 4+3}{8} \\ &= \frac{35}{8} \end{aligned}$$

20. $$\begin{aligned} 15\frac{3}{4} &= \frac{4\times 15+3}{4} \\ &= \frac{63}{4} \end{aligned}$$

21. $$\begin{aligned} 5\frac{2}{7} &= \frac{7\times 5+2}{7} \\ &= \frac{37}{7} \end{aligned}$$

22. $$\begin{aligned} 6\frac{3}{5} &= \frac{5\times 6+3}{5} \\ &= \frac{33}{5} \end{aligned}$$

23. $$\begin{array}{r} 5 \\ 8\overline{)45} \\ \underline{40} \\ 5 \end{array}$$

$$\frac{45}{8} = 5\frac{5}{8}$$

24. $$\begin{array}{r} 4 \\ 21\overline{)100} \\ \underline{84} \\ 16 \end{array}$$

$$\frac{100}{21} = 4\frac{16}{21}$$

25. $7\overline{)23}$ quotient 3
$\underline{21}$
2

$\frac{23}{7} = 3\frac{2}{7}$

26. $9\overline{)74}$ quotient 8
$\underline{72}$
2

$\frac{74}{9} = 8\frac{2}{9}$

27. $3\frac{15}{55} = 3\frac{5 \times 3}{5 \times 11} = 3\frac{3}{11}$

28. $\frac{234}{16} = \frac{117 \times 2}{8 \times 2} = \frac{117}{8}$

29. $32\overline{)132}$ quotient 4
$\underline{128}$
4

$\frac{132}{32} = 4\frac{1}{8}$

30. $\frac{4}{7} \times \frac{5}{11} = \frac{4 \times 5}{7 \times 11}$
$= \frac{20}{77}$

31. $\frac{7}{9} \times \frac{21}{35} = \frac{1}{3} \times \frac{7}{5}$
$= \frac{7}{15}$

32. $12 \times \frac{3}{7} \times 0 = 0$

33. $\frac{3}{5} \times \frac{2}{7} \times \frac{10}{27} = \frac{1}{1} \times \frac{2}{7} \times \frac{2}{9} = \frac{4}{63}$

34. $12 \times 8\frac{1}{5} = \frac{12}{1} \times \frac{41}{5}$
$= \frac{492}{5}$
$= 98\frac{2}{5}$

35. $5\frac{1}{4} \times 4\frac{6}{7} = \frac{21}{4} \times \frac{34}{7}$
$= \frac{3}{2} \times \frac{17}{1} = \frac{51}{2}$
$= 25\frac{1}{2}$

36. $5\frac{1}{8} \times 3\frac{1}{5} = \frac{41}{8} \times \frac{16}{5}$
$= \frac{82}{5}$ or $16\frac{2}{5}$

37. $35 \times \frac{7}{10} = \frac{35}{1} \times \frac{7}{10}$
$= \frac{7}{1} \times \frac{7}{2} = \frac{49}{2}$
$= 24\frac{1}{2}$

38. $37\frac{5}{8} \times 18 = \frac{301}{8} \cdot \frac{18}{1}$
$= \frac{301}{4} \times \frac{9}{1}$
$= \frac{2709}{4} = 677\frac{1}{4}$

$\$677\frac{1}{4}$

39. $12\frac{1}{5} \times 3\frac{4}{7} = \frac{61}{5} \times \frac{25}{7}$
$= \frac{305}{7}$
$= 43\frac{4}{7}$

The area is $43\frac{4}{7}$ square inches.

40. $\frac{3}{7} \div \frac{2}{5} = \frac{3}{7} \times \frac{5}{2}$
$= \frac{15}{14} = 1\frac{1}{14}$

41. $\frac{3}{5} \div \frac{1}{10} = \frac{3}{5} \times \frac{10}{1}$
$= 6$

42. $1200 \div \frac{5}{8} = \frac{1200}{1} \times \frac{8}{5}$
$= 1920$

43. $900 \div \frac{3}{5} = \frac{900}{1} \times \frac{5}{3}$
$= 1500$

44. $5\frac{3}{4} \div 11\frac{1}{2} = \frac{23}{4} \div \frac{23}{2}$
$= \frac{23}{4} \times \frac{2}{23}$
$= \frac{1}{2}$

45. $20 \div 2\frac{1}{2} = \frac{20}{1} \div \frac{5}{2}$
$= \frac{20}{1} \times \frac{2}{5}$
$= 8$

46. $0 \div 3\frac{7}{5} = 0$

47. $4\frac{2}{11} \div 3 = \frac{46}{11} \div \frac{3}{1}$
$= \frac{46}{11} \times \frac{1}{3}$
$= \frac{46}{33}$
$= 1\frac{13}{33}$

48. $342 \div 28\frac{1}{2} = \frac{342}{1} \div \frac{57}{2}$
$= \frac{342}{1} \times \frac{2}{57}$
$= 6 \times 2$
$= 12$

12 rolls are needed.

49. $420 \div 2\frac{1}{4} = \frac{420}{1} \div \frac{9}{4}$
$= \frac{420}{1} \times \frac{4}{9}$
$= \frac{140}{1} \times \frac{4}{3} = \frac{560}{3}$
$= 186\frac{2}{3}$ calories

50. $14 = 2 \times 7$
$49 = 7 \times 7$
$\text{LCD} = 2 \times 7 \times 7 = 98$

51. $40 = 2 \times 2 \times 2 \times 5$
$30 = 2 \times 3 \times 5$
$\text{LCD} = 2 \times 2 \times 2 \times 3 \times 5 = 120$

52. $18 = 2 \times 3 \times 3$
$6 = 2 \times 3$
$45 = 3 \times 3 \times 5$
$\text{LCD} = 2 \times 3 \times 3 \times 5 = 90$

53. $\frac{3}{7} = \frac{3}{7} \times \frac{8}{8}$
$= \frac{24}{56}$

54. $\frac{11}{24} = \frac{11}{24} \times \frac{3}{3}$
$= \frac{33}{72}$

55. $\frac{9}{43} = \frac{9}{43} \times \frac{4}{4}$
$= \frac{36}{172}$

56. $\frac{17}{18}=\frac{17}{18}\times\frac{11}{11}$
$=\frac{187}{198}$

57. $\frac{3}{7}-\frac{5}{14}=\frac{3}{7}\times\frac{2}{2}-\frac{15}{14}$
$=\frac{6}{14}-\frac{5}{14}$
$=\frac{1}{14}$

58. $\frac{1}{2}+\frac{1}{3}+\frac{1}{4}=\frac{1}{2}\times\frac{6}{6}+\frac{1}{3}\times\frac{4}{4}+\frac{1}{4}\times\frac{3}{3}$
$=\frac{6}{12}+\frac{4}{12}+\frac{3}{12}$
$=\frac{13}{12}=1\frac{1}{12}$

59. $\frac{4}{7}+\frac{7}{9}=\frac{4}{7}\times\frac{9}{9}+\frac{7}{9}\times\frac{7}{7}$
$=\frac{36}{63}+\frac{49}{63}$
$=\frac{85}{63}=1\frac{22}{63}$

60. $\frac{7}{8}-\frac{3}{5}=\frac{7}{8}\times\frac{5}{5}-\frac{3}{5}\times\frac{8}{8}$
$=\frac{35}{40}-\frac{24}{40}$
$=\frac{11}{40}$

61. $\frac{7}{30}+\frac{2}{21}=\frac{7}{30}\times\frac{7}{7}+\frac{2}{21}\times\frac{10}{10}$
$=\frac{49}{210}+\frac{20}{210}$
$=\frac{69}{210}=\frac{23}{70}$

62. $\frac{5}{18}+\frac{5}{12}=\frac{5}{18}\times\frac{2}{2}+\frac{5}{12}\times\frac{3}{3}$
$=\frac{10+15}{36}$
$=\frac{25}{36}$

63. $\frac{15}{16}-\frac{13}{24}=\frac{15}{16}\times\frac{3}{3}-\frac{13}{24}\times\frac{2}{2}$
$=\frac{45}{48}-\frac{26}{48}$
$=\frac{19}{48}$

64. $\frac{14}{15}-\frac{3}{25}=\frac{14}{15}\times\frac{5}{5}-\frac{3}{25}\times\frac{3}{3}$
$=\frac{70}{75}-\frac{9}{75}$
$=\frac{61}{75}$

65. $2-\frac{3}{4}=\frac{2}{1}-\frac{3}{4}$
$=\frac{8}{4}-\frac{3}{4}$
$=\frac{5}{4}$ or $1\frac{1}{4}$

66. $6-\frac{5}{9}=5\frac{9}{9}-\frac{5}{9}$
$=5\frac{4}{9}$

67. $3+5\frac{2}{3}=8\frac{2}{3}$

68. $8+12\frac{5}{7}=20\frac{5}{7}$

69.

$$\begin{array}{r} 3\frac{3}{8} \\ +\ 2\frac{3}{4} \\ \hline \end{array} \qquad \begin{array}{r} 3\frac{3}{8} \\ +\ 2\frac{6}{8} \\ \hline 5\frac{9}{8}=6\frac{1}{8} \end{array}$$

70.
$$\begin{array}{r} 5\frac{11}{16} \\ -\ 2\frac{1}{5} \\ \hline \end{array} \qquad \begin{array}{r} 5\frac{55}{80} \\ -\ 2\frac{16}{80} \\ \hline 3\frac{39}{80} \end{array}$$

71.
$$\frac{3}{5}\times\frac{1}{2}+\frac{2}{5}\div\frac{2}{3}=\frac{3}{5}\times\frac{1}{2}+\frac{2}{5}\times\frac{3}{2}$$
$$=\frac{3}{10}+\frac{3}{5}$$
$$=\frac{3}{10}+\frac{3}{5}\times\frac{3}{2}$$
$$=\frac{3}{10}+\frac{6}{10}$$
$$=\frac{9}{10}$$

72.
$$\left(\frac{3}{7}-\frac{1}{14}\right)\times\frac{8}{9}=\left(\frac{6}{14}-\frac{1}{14}\right)\times\frac{8}{9}$$
$$=\frac{5}{14}\times\frac{8}{9}$$
$$=\frac{40}{126}=\frac{20}{63}$$

73.
$$1\frac{7}{8}+2\frac{3}{4}+4\frac{1}{10}$$
$$=1\frac{70}{80}+2\frac{60}{80}+4\frac{8}{80}$$
$$=7\frac{138}{80}$$
$$=8\frac{58}{80}$$
$$=8\frac{29}{40}\text{ miles}$$

74.
$$\begin{array}{r} 28\frac{1}{6} \\ -\ 1\frac{5}{6} \\ \hline \end{array} \qquad \begin{array}{r} 27\frac{7}{6} \\ -\ 1\frac{5}{6} \\ \hline 26\frac{2}{6}=26\frac{1}{3} \end{array}$$

Then: $26\frac{1}{3}\times10\frac{3}{4}=\frac{79}{3}\times\frac{43}{4}$
$$=\frac{3397}{12}$$
$$=283\frac{1}{12}\text{ miles}$$

75.
$$3\frac{1}{3}\times2\frac{1}{2}=\frac{10}{3}\times\frac{1}{2}=\frac{5}{3}=1\frac{2}{3}\text{ cups sugar}$$
$$4\frac{1}{4}\times\frac{1}{2}=\frac{17}{4}\times\frac{1}{2}=\frac{17}{8}=2\frac{1}{8}\text{ cups sugar}$$

76.
$$24\frac{1}{4}\times8\frac{1}{2}=\frac{97}{4}\times\frac{17}{2}$$
$$=\frac{1649}{8}$$
$$=206\frac{1}{8}\text{ miles}$$

77.
$$48\div3\frac{1}{5}=\frac{48}{1}\div\frac{16}{5}$$
$$=\frac{48}{1}\times\frac{5}{16}=\frac{3}{1}\times\frac{5}{1}$$
$$=15\text{ lengths}$$

78.
$$15\frac{3}{4}-6\frac{1}{8}=15\frac{6}{8}-\frac{1}{8}$$
$$=9\frac{5}{8}\text{ liters}$$

79.
$$\begin{array}{r} 12 \\ 9 \\ +\ 14 \\ \hline 35 \end{array}$$
$$35\div5=7$$
$$7\times32\frac{1}{2}=\frac{7}{1}\times\frac{65}{2}$$
$$=\frac{455}{2}$$
$$=227\frac{1}{2}$$

It will take $227\frac{1}{2}$ minutes or

3 hours and $47\frac{1}{2}$ minutes.

80. Regular pay: $4\frac{1}{2}\times 8=\frac{9}{2}\times\frac{8}{1}$

$=36$

Overtime rate: $1\frac{1}{2}\times 4\frac{1}{2}=\frac{3}{2}\times\frac{9}{2}$

$=\frac{27}{4}$

Overtime pay: $\frac{27}{3}\times 3=\frac{27}{4}\times\frac{3}{1}$

$=\frac{81}{4}=20\frac{1}{4}$

Total pay: $36+20\frac{1}{4}=56\frac{1}{4}$

$\$56\frac{1}{4}$

81. $80\times 12\frac{1}{4}=\frac{80}{1}\times\frac{49}{24}=980$

$80\times 18=1440$

$$\begin{array}{r} 1440 \\ -\ 980 \\ \hline 460 \end{array}$$

She made \$460.

82. $1\frac{1}{2}+\frac{1}{16}+\frac{1}{8}+\frac{1}{4}=1\frac{8}{16}+\frac{1}{16}+\frac{2}{16}+\frac{4}{16}$

$=1\frac{15}{16}$

$3-1\frac{15}{16}=2\frac{16}{16}-1\frac{15}{16}$

$=1\frac{1}{16}$ inch

83.

$$\begin{array}{rcr} \frac{1}{10}\times 880 & = & 98 \\ \frac{1}{2}\times 880 & = & 440 \\ +\ \frac{1}{8}\times 880 & = & +110 \\ \hline & & 638 \end{array}$$

Left over:

$$\begin{array}{r} 880 \\ -638 \\ \hline \$242 \end{array}$$

84. A. $460\div 18\frac{2}{5}=\frac{460}{1}\div\frac{92}{5}$

$=\frac{460}{1}\times\frac{5}{92}=25$

25 miles per gallon.

B. $18\frac{2}{5}\times 1\frac{1}{5}=\frac{92}{5}\times\frac{6}{5}$

$=\frac{552}{25}=22\frac{2}{25}$

$\$22\frac{2}{25}$

85. $\frac{27}{63}=\frac{27\div 9}{63\div 9}=\frac{3}{7}$

86. $\frac{7}{15}+\frac{11}{25}=\frac{35}{75}+\frac{33}{75}=\frac{68}{75}$

87.

$$\begin{array}{r} 4\frac{1}{3} \\ -\ 2\frac{11}{12} \\ \hline \end{array}\qquad \begin{array}{r} 4\frac{4}{12} \\ -\ 2\frac{11}{12} \\ \hline \end{array}\qquad \begin{array}{r} 3\frac{16}{12} \\ -\ 2\frac{11}{12} \\ \hline 1\frac{5}{12} \end{array}$$

88. $\frac{36}{49}\times\frac{14}{33}=\frac{3\times 12\times 2\times 7}{3\times 11\times 7\times 7}=\frac{24}{77}$

89. $5\frac{2}{3}\div 1\frac{1}{6}=\frac{17}{3}\div\frac{7}{6}$

$=\frac{17}{3}\times\frac{6}{7}$

$=\frac{34}{7}$ or $4\frac{6}{7}$

90. $\left(\frac{4}{7}\right)^3=\frac{4}{7}\times\frac{4}{7}\times\frac{4}{7}=\frac{64}{343}$

91.

$$\begin{array}{r} 3\frac{7}{8} \\ +\ 2\frac{5}{16} \\ \hline \end{array}\qquad \begin{array}{r} 3\frac{14}{16} \\ +\ 2\frac{5}{16} \\ \hline 5\frac{19}{16}=6\frac{3}{16} \end{array}$$

92. $5\frac{1}{2}\times 7\frac{5}{6}=\frac{11}{2}\times\frac{47}{6}=\frac{517}{12}=43\frac{1}{12}$

93. $120 \div 3\frac{3}{5} = \frac{120}{1} \div \frac{18}{5}$
$= \frac{120}{1} \times \frac{5}{18}$
$= \frac{6 \times 20}{1} \times \frac{5}{6 \cdot 3}$
$= \frac{100}{3}$ or $33\frac{1}{3}$

How Am I Doing? Chapter 2 Test

1. $\frac{3}{5}$; 3 of the 5 parts are shaded.

2. $\frac{311}{388}$

3. $\frac{18}{42} = \frac{18 \div 6}{42 \div 6}$
$= \frac{3}{7}$

4. $\frac{15}{70} = \frac{15 \div 5}{70 \div 5}$
$= \frac{3}{14}$

5. $\frac{225}{50} = \frac{225 \div 25}{50 \div 25}$
$= \frac{9}{2}$

6. $6\frac{4}{5} = \frac{6 \times 5 + 4}{5}$
$= \frac{34}{5}$

7. $14\overline{)145}$ quotient 10, -14, remainder 5. $\frac{145}{14} = 10\frac{5}{14}$

8. $42 \times \frac{2}{7} = \frac{42}{1} \times \frac{2}{7} = \frac{6 \times 7 \times 2}{1 \times 7}$
$= \frac{12}{1} = 12$

9. $\frac{7}{9} \times \frac{2}{5} = \frac{7 \times 2}{9 \times 5}$
$= \frac{14}{45}$

10. $2\frac{2}{3} \times 5\frac{1}{4} = \frac{8}{3} \times \frac{21}{4}$
$= \frac{2 \times 4 \times 3 \times 7}{3 \times 4}$
$= 14$

11. $\frac{7}{8} \div \frac{5}{11} = \frac{7}{8} \times \frac{11}{5}$
$= \frac{7 \times 11}{8 \times 5} = \frac{77}{40}$
$= 1\frac{37}{40}$

12. $\frac{12}{31} \div \frac{8}{13} = \frac{12}{31} \times \frac{13}{8} = \frac{3 \times 4 \times 13}{31 \times 2 \times 4} = \frac{39}{62}$

13. $7\frac{1}{5} \div 1\frac{1}{25} = \frac{36}{5} \div \frac{26}{25} = \frac{36}{5} \times \frac{25}{26}$
$= \frac{2 \times 18 \times 5 \times 5}{5 \times 2 \times 13} = \frac{18 \times 5}{13}$
$= \frac{90}{13} = 6\frac{12}{13}$

14. $5\frac{1}{7} \div 3 = \frac{36}{7} \div \frac{3}{1} = \frac{36}{7} \times \frac{1}{3}$
$= \frac{3 \times 12 \times 1}{7 \times 3}$
$= \frac{12}{7} = 1\frac{5}{7}$

15. $12 = 2 \times 2 \times 3$
$18 = 2 \times 3 \times 3$
$\text{LCD} = 2^2 \times 3^2 = 36$

16. $16 = 2 \times 2 \times 2 \times 2$
$24 = 2 \times 2 \times 2 \times 3$
$\text{LCD} = 2 \times 2 \times 2 \times 2 \times 3 = 48$

17.
$$4 = 2\times 2$$
$$8 = 2\times 2\times 2$$
$$6 = 2\times 3$$
$$\text{LCD} = 2\times 2\times 2\times 3 = 24$$

18. $\frac{5}{12} = \frac{5}{12}\times\frac{6}{6} = \frac{30}{72}$

19. $\frac{7}{9} - \frac{5}{12} = \frac{28}{36} - \frac{15}{36} = \frac{13}{36}$

20.
$$\frac{2}{15} + \frac{5}{12} = \frac{8}{60} + \frac{25}{60} = \frac{33}{60} = \frac{11}{20}$$

21. $\frac{1}{4} + \frac{3}{7} + \frac{3}{14} = \frac{7}{28} + \frac{12}{28} + \frac{6}{28} = \frac{25}{28}$

22.
$$8\frac{3}{5} + 5\frac{4}{7} = 8\frac{21}{35} + 5\frac{20}{35} = 13\frac{41}{35} = 14\frac{6}{35}$$

23.
$$18\frac{6}{7} - 13\frac{13}{14} = 18\frac{12}{14} - 13\frac{13}{14} = 17\frac{26}{14} - 13\frac{13}{14} = 4\frac{13}{14}$$

24. $\frac{2}{9} \div \frac{8}{3} \times \frac{1}{4} = \frac{2}{9}\times\frac{3}{8}\times\frac{1}{4} = \frac{1}{48}$

25.
$$\left(\frac{1}{2}+\frac{1}{3}\right)\times\frac{7}{5} = \left(\frac{3}{6}+\frac{2}{6}\right)\times\frac{7}{5} = \frac{5}{6}\times\frac{7}{5} = \frac{7}{6} = 1\frac{1}{6}$$

26. $16\frac{1}{2}\times 9\frac{1}{3} = \frac{33}{2}\times\frac{28}{3} = 11\times 14 = 154$

The kitchen is 154 square feet.

27.
$$18\frac{2}{3} \div 2\frac{1}{3} = \frac{56}{3} \div \frac{7}{3} = \frac{56}{3}\times\frac{3}{7} = \frac{8\times 7\times 3}{3\times 7} = 8$$

He can make 8 packages.

28. $\frac{9}{10} - \frac{1}{5} = \frac{9}{10} - \frac{2}{10} = \frac{7}{10}$

He has $\frac{7}{10}$ of a mile left to walk.

29.
$$4\frac{1}{8} + 3\frac{1}{6} + 6\frac{3}{4} = 4\frac{3}{24} + 3\frac{4}{24} + 6\frac{18}{24} = 13\frac{25}{24} = 14\frac{1}{24}$$

She jogged $14\frac{1}{24}$ miles.

30. A.
$$\frac{1}{4}\times 120 = \frac{1}{4}\times\frac{120}{1} = 30$$
$$\frac{1}{12}\times 120 = \frac{1}{12}\times\frac{120}{1} = 10$$
$$\frac{1}{3}\times 120 = \frac{1}{3}\times\frac{120}{1} = 40$$
$$120 - 30 - 10 - 40 = 120 - 80 = 40$$

They shipped 40 oranges.

B. $24\times 40 = 960$

It cost 960 cents or $9.60.

31. A.
$$48\frac{1}{8} \div \frac{5}{8} = \frac{385}{8}\times\frac{8}{5} = \frac{385}{5} = 77 \text{ candles}$$

B. Cost is

$$48\frac{1}{8}\times 2\div 77=\frac{385}{8}\times\frac{2}{1}\times\frac{1}{77}$$
$$=\frac{385}{308}=1\frac{77}{308}$$
$$=\$1\frac{1}{4}$$

C. Profit per candle:

$$12-1\frac{1}{4}=11\frac{4}{4}-1\frac{1}{4}$$
$$=10\frac{3}{4}$$

Profit is

$$10\frac{3}{4}\times 77=\frac{43}{4}\times\frac{77}{1}$$
$$=\frac{3311}{4}=\$827\frac{3}{4}$$

Cumulative Test for Chapters 1-2

1. 84,361,208 = Eighty-four million, three hundred sixty-one thousand, two hundred eight.

2.
```
   128
   452
   178
    34
+   77
------
   869
```

3.
```
  156,200
  364,700
+ 198,320
---------
  719,220
```

4.
```
 5718
-3643
-----
 2075
```

5.
```
  1,000,361
-   983,145
-----------
     17,216
```

6.
```
  126
×  38
-----
 1008
 378
-----
 4788
```

7.
```
  16,908
×     12
--------
  33,816
 16908
--------
 202,896
```

8.
```
     4658
 7)32,606
   28
    46
    42
     40
     35
      56
      56
       0
```

9.
```
     308
 15)4631
    45
     131
     120
      11
```

10. $7^2=7\times 7=49$

11. 6,037,452 rounds to 6,037,000.

12.
$$6\times 2^3+12\div(4+2)=6\times 2^3+12\div 6$$
$$=6\times 8+12\div 6$$
$$=48+2=50$$

13. $3\times\$26+2\times\$48=\$174$

14. $516+199+203=918$ for checks

$64+1160-918=1224-918=306$

Her balance will be $306.

15.

112	Women	Men
−83	$\frac{83}{112}$	$\frac{29}{112}$
29		

16. $\frac{28}{52} = \frac{28 \div 4}{52 \div 4} = \frac{7}{13}$

17. $18\frac{3}{4} = \frac{4 \times 18 + 3}{4} = \frac{75}{4}$

18. $7\overline{)100}$ quotient 14; -7, 30, -28, remainder 2

$\frac{100}{7} = 14\frac{2}{7}$

19. $3\frac{7}{8} \times 2\frac{5}{6} = \frac{31}{8} \times \frac{17}{6}$

$= \frac{527}{48} = 10\frac{47}{48}$

20. $\frac{44}{49} \div 2\frac{13}{21} = \frac{44}{49} \div \frac{55}{21} = \frac{44}{49} \times \frac{21}{55}$

$= \frac{4 \times 11 \times 3 \times 7}{7 \times 7 \times 5 \times 11} = \frac{12}{35}$

21. $8 = 2 \times 2 \times 2$

$10 = 2 \times 5$

$\text{LCD} = 2^3 \times 5 = 40$

22. $\frac{7}{18} + \frac{20}{27} = \frac{21}{54} + \frac{40}{54} = \frac{61}{54}$ or $1\frac{7}{54}$

23. $2\frac{1}{8} + 6\frac{3}{4} = 2\frac{1}{8} + 6\frac{6}{8} = 8\frac{7}{8}$

24. $12\frac{1}{5} - 4\frac{2}{3} = 12\frac{3}{15} - 4\frac{10}{15}$

$= 11\frac{18}{15} - 4\frac{10}{15} = 7\frac{8}{15}$

25. $\frac{1}{3} + \frac{5}{8} \div \frac{5}{4} = \frac{1}{3} + \frac{5}{8} \times \frac{4}{5}$

$= \frac{1}{3} + \frac{1}{2} = \frac{2}{6} + \frac{3}{6}$

$= \frac{5}{6}$

26.

$$\begin{array}{r} 5\frac{1}{2} \\ +\ 6\frac{3}{4} \\ \hline \end{array} \qquad \begin{array}{r} 5\frac{2}{4} \\ +\ 6\frac{3}{4} \\ \hline 11\frac{5}{4} = 12\frac{1}{4} \end{array}$$

He must lose $2\frac{3}{4}$ pounds.

27. $221\frac{2}{5} \div 9 = \frac{1107}{5} \div \frac{9}{1}$

$= \frac{1107}{5} \times \frac{1}{9} = \frac{123}{5}$

$= 24\frac{3}{5}$ miles per gallon

28. $2\frac{1}{2} \times 3\frac{1}{4} = \frac{5}{2} \times \frac{13}{4} = \frac{65}{8} = 8\frac{1}{8}$

$2\frac{1}{2} \times 2\frac{1}{3} = \frac{5}{2} \times \frac{7}{3} = \frac{35}{6} = 5\frac{5}{6}$

She needs $8\frac{1}{8}$ cups of sugar and $5\frac{5}{6}$ cups of flour.

29.

$$\begin{array}{r} 30,000 \\ \times\ \ 2,000 \\ \hline 60,000,000 \end{array}$$ miles

30. $960 \times \frac{1}{6} = \frac{960}{1} \times \frac{1}{6} = \frac{960}{6} = 160$

They cost $160.

Chapter 3

3.1 Exercises

1. A decimal fraction is a fraction whose denominator is a power of 10. $\frac{23}{100}$ and $\frac{563}{1000}$ are decimal fractions.

3. Hundred-thousandths

5. 0.57 = Fifty-seven hundredths

7. 3.8 = Three and eight tenths

9. 5.803 = Five and eight hundred three thousandths

11. 28.0037 = Twenty-eight and thirty-seven ten-thousandths

13. \$124.20 = One hundred twenty-four and $\frac{20}{100}$ dollars

15. \$1236.08 = One thousand two hundred thirty-six and $\frac{8}{100}$ dollars

17. \$12,015.45 = twelve thousand fifteen and $\frac{45}{100}$ dollars

19. seven tenths = 0.7

21. twelve hundredths 0.12

23. four hundred eighty-one thousandths = 0.481

25. two hundred eighty-six millionths = 0.000286

27. $\frac{7}{10} = 0.7$

29. $\frac{76}{100} = 0.76$

31. $\frac{1}{100} = 0.01$

33. $\frac{53}{1000} = 0.053$

35. $\frac{2403}{10,000} = 0.2403$

37. $8\frac{3}{10} = 8.3$

39. $84\frac{13}{100} = 84.13$

41. $3\frac{529}{1000} = 3.529$

43. $126\frac{571}{10,000} = 126.0571$

45. $0.02 = \frac{2}{100} = \frac{1}{50}$

47. $3.6 = 3\frac{6}{10} = 3\frac{3}{5}$

49. $7.41 = 7\frac{41}{100}$

51. $12.625 = 12\frac{625}{1000} = 12\frac{5}{8}$

53. $7.0615 = 7\frac{615}{10,000} = 7\frac{123}{2000}$

55. $8.0108 = 8\frac{108}{10,000} = 8\frac{27}{2500}$

57. $235.1254 = 235\frac{1254}{10,000} = 235\frac{627}{5000}$

59. $0.0187 = \dfrac{187}{10,000}$

61. a. $\dfrac{31,700}{100,000} = \dfrac{317}{1000}$

b. $\dfrac{30,100}{100,000} = \dfrac{301}{1000}$

63. $\dfrac{4}{1,000,000} = \dfrac{1}{250,000}$

Cumulative Review

65.
$$\begin{array}{r} 207 \\ 54 \\ 123 \\ 86 \\ +\ 55 \\ \hline 525 \end{array}$$

67. 56,800

3.2 Exercises

1. 1.3 > 1.29

3. 0.34 = 0.340

5. 18.92 < 18.93

7. 0.006 > 0.00063

9. 1.002 < 1.0021

11. 126.34 > 125.35

13. 0.888 < 0.8888

15. 0.777 > 0.7077

17. $\dfrac{72}{1000} = 0.072$

19. $\dfrac{8}{10} = 0.8$

$\dfrac{8}{10} > 0.08$

21. 12.6, 12.65, 12.8

23. 0.007, 0.0071, 0.05

25. 5.12, 5.2, 5.23, 5.3

27. 26.003, 26.033, 26.034, 26.04

29. 18.006, 18.060, 18.065, 18.066, 18.606

31. 5.67 rounds to 5.7

33. 28.98 rounds to 29.0

35. 578.064 rounds to 578.1

37. 2176.83 rounds to 2176.8

39. 26.032 rounds to 26.03

41. 36.997 rounds to 37.00

43. 156.1749 rounds to 156.17

45. 2786.706 rounds to 2786.71

47. 1.06132 rounds to 1/061

49. 0.05951 rounds to 0.0595

51. 5.00761238 rounds to 5.00761

53. 135.564 rounds to 136

55. $2536.85 rounds to $2537

57. $15,020.50 rounds to $15, 021

59. $56.9832 rounds to $56.98

61. $5783.716 rounds to $5783.72

63. 0.60869 rounds to 0.609
0.52127 rounds to 0.521

65. 365.24122 rounds to 365.24

67. $\frac{6}{100} = 0.06$ and $\frac{6}{10} = 0.6$

0.0059, 0.006, 0.519, $\frac{6}{100}$, 0.0601, 0.0612,

0.062, $\frac{6}{10}$, 0.61

69. You should consider only one digit to the right of the decmal place that you wish to round to. 86.23498 is closer to 86.23 than to 86.24.

Cumulative Review

71.
$$\begin{array}{r} 3\tfrac{1}{4} \\ 2\tfrac{1}{2} \\ +\ 6\tfrac{3}{8} \\ \hline \end{array} \qquad \begin{array}{r} 3\tfrac{2}{8} \\ 2\tfrac{4}{8} \\ +\ 6\tfrac{3}{8} \\ \hline 11\tfrac{9}{8} = 12\tfrac{1}{8} \end{array}$$

73.
$$\begin{array}{r} 47{,}073 \\ -\ 46{,}381 \\ \hline 692 \text{ miles} \end{array}$$

3.3 Exercises

1.
$$\begin{array}{r} 57.1 \\ +19.7 \\ \hline 76.8 \end{array}$$

3.
$$\begin{array}{r} 718.98 \\ +496.57 \\ \hline 1215.55 \end{array}$$

5.
$$\begin{array}{r} 13.4 \\ 7.6 \\ +\ 275.2 \\ \hline 296.2 \end{array}$$

7.
$$\begin{array}{r} 4.71 \\ +\ 8.05 \\ \hline 12.76 \end{array}$$

9.
$$\begin{array}{r} 4.9637 \\ 28.1200 \\ +\ 3.6450 \\ \hline 36{,}7287 \end{array}$$

11.
$$\begin{array}{r} 12.00 \\ 3.62 \\ +\ 51.80 \\ \hline 67.42 \end{array}$$

13.
$$\begin{array}{r} 156.35 \\ 2.79 \\ 126.30 \\ +\ 86.00 \\ \hline 371.44 \end{array}$$

15.
$$\begin{array}{r} 753.61 \\ 28.75 \\ 162.30 \\ 100.50 \\ +\ 67.05 \\ \hline 1112.21 \end{array}$$

17. Perimeter:
$$\begin{array}{r} 5.26 \\ 9.28 \\ +\ 6.50 \\ \hline 21.04 \text{ ft} \end{array}$$

19.
$$\begin{array}{r} 1.75 \\ 2.50 \\ 1.55 \\ +\ 2.80 \\ \hline 8.60 \text{ or } 8.6 \text{ pounds} \end{array}$$

21.
$$\begin{array}{r} 4.99 \\ 12.50 \\ 11.85 \\ 28.50 \\ 3.29 \\ +\ 16.99 \\ \hline \$78.12 \end{array}$$

23.
$$\begin{array}{r} 23{,}276.0 \\ +\quad 778.9 \\ \hline 47{,}054.9 \end{array}$$ miles

25.
$$\begin{array}{r} 18.42 \\ 706.15 \\ 21.03 \\ 45.00 \\ +\ 621.37 \\ \hline \$1411.97 \end{array}$$ total

27.
$$\begin{array}{r} 6.8 \\ -\ 2.9 \\ \hline 3.9 \end{array}$$

29.
$$\begin{array}{r} 35.75 \\ -\ 9.82 \\ \hline 25.93 \end{array}$$

31.
$$\begin{array}{r} 126.00 \\ -\ 76.22 \\ \hline 49.78 \end{array}$$

33.
$$\begin{array}{r} 586.513 \\ -\ 78.200 \\ \hline 508.313 \end{array}$$

35.
$$\begin{array}{r} 162.40 \\ -\ 97.52 \\ \hline 64.88 \end{array}$$

37.
$$\begin{array}{r} 24.0079 \\ -\ 19.3614 \\ \hline 4.6465 \end{array}$$

39.
$$\begin{array}{r} 8.000 \\ -\ 1.263 \\ \hline 6.737 \end{array}$$

41.
$$\begin{array}{r} 7362.14 \\ -\ 6173.07 \\ \hline 1189.07 \end{array}$$

43.
$$\begin{array}{r} 1.5000 \\ -\ 0.0365 \\ \hline 1.4635 \end{array}$$

45.
$$\begin{array}{r} 123.621 \\ +\ 52.960 \\ \hline 176.581 \end{array}$$

47.
$$\begin{array}{r} 98.30 \\ -\ 56.71 \\ \hline 41.59 \end{array}$$

49.
$$\begin{array}{r} 0.0763 \\ 2.0000 \\ +\ 3.1600 \\ \hline 5.2363 \end{array}$$

51.
$$\begin{array}{r} 197.600 \\ -124.375 \\ \hline 73.225 \end{array}$$

53.
$$\begin{array}{r} 3.264 \\ -\ 1.800 \\ \hline 1.464 \end{array}$$ meters

55.
$$\begin{array}{r} 37{,}026.65 \\ -\qquad 79.49 \\ \hline \$36{,}947.16 \end{array}$$

57.
$$\begin{array}{r} 47.70 \\ +\ 7.00 \\ \hline 54.70 \end{array} \qquad \begin{array}{r} 100.00 \\ -\ 54.70 \\ \hline 45.30 \end{array}$$
$45.30 change

59.
$$\begin{array}{r} 12.62 \\ -\ 0.98 \\ \hline 11.64 \end{array}$$ cm

61. $2.45+1.35-0.85$
$=3.80-0.85$
$=2.95$ liters

63.
$$\begin{array}{r} 0.0150 \\ -\ 0.0089 \\ \hline 0.0061 \end{array}$$ milligrams

65.
$$\begin{array}{r} 43.8 \\ -\ 37.6 \\ \hline \end{array}$$
$6.2 billion or
$6,200,000,000

67.
$$\begin{array}{r} 271.7 \\ -\ 109.6 \\ \hline \$162.1 \end{array}$$
$162.1 billion = $162,100,000,000

69.
$$\begin{array}{r} 2.60 \\ 1.50 \\ 1.30 \\ 0.80 \\ +\ 2.20 \\ \hline \$8.40 \end{array}$$
Exact:
$$\begin{array}{r} 2.63 \\ 1.47 \\ 1.26 \\ 0.79 \\ +\ 2.19 \\ \hline \$8.34 \end{array}$$
Estimate is close to actual amount.
Difference is 6 cents.

71. $x+7.1=15.5$
$$\begin{array}{r} 15.5 \\ -\ 7.1 \\ \hline 8.4 \end{array}$$
$x=8.4$

73. $156.9+x=200.6$
$$\begin{array}{r} 200.6 \\ -\ 156.9 \\ \hline 43.7 \end{array}$$
$x=43.7$

75. $4.162=x+2.053$
$$\begin{array}{r} 4.162 \\ -\ 2.053 \\ \hline 2.109 \end{array}$$
$x=2.109$

Cumulative Review

77.
$$\begin{array}{r} 2536 \\ \times\ 8 \\ \hline 20,288 \end{array}$$

79. $\frac{22}{7}\times\frac{49}{50}=\frac{77}{25}=3\frac{2}{25}$

3.4 Exercises

1. Each factor has two decimal places. You add the number of decimal places to get four decimal places. You multiply 67×8 to obtain 536. Now you must place the decimal point four places to the left in your answer. The result is 0.0536.

3. When you multiply a number by 100 you move the decimal point two places to the right. The answer would be 0.78.

5.
$$\begin{array}{r} 0.6 \\ \times\ 0.2 \\ \hline 0.12 \end{array}$$

7.
$$\begin{array}{r} 0.12 \\ \times\ \ 0.5 \\ \hline 0.060 = 0.06 \end{array}$$

9.
$$\begin{array}{r} 0.0036 \\ \times\ \ \ \ 0.8 \\ \hline 0.00288 \end{array}$$

11.
$$\begin{array}{r} 452 \\ \times\ 0.12 \\ \hline 904 \\ 452\ \\ \hline 5424 \end{array}$$

13.
$$\begin{array}{r} 0.043 \\ \times\ 0.012 \\ \hline 0086 \\ 0043\ \\ \hline 0.000516 \end{array}$$

15.
$$\begin{array}{r} 10.97 \\ \times\ \ 0.06 \\ \hline 0.6582 \end{array}$$

17.
$$\begin{array}{r} 5167 \\ \times\ \ 0.19 \\ \hline 46503 \\ 5167\ \\ \hline 981.73 \end{array}$$

19.
$$\begin{array}{r} 2.163 \\ \times\ 0.008 \\ \hline 0.017304 \end{array}$$

21.
$$\begin{array}{r} 0.7613 \\ \times\ \ 1009 \\ \hline 68517 \\ 761300\ \\ \hline 768.1517 \end{array}$$

23.
$$\begin{array}{r} 2350 \\ \times\ \ 3.6 \\ \hline 14100 \\ 7050\ \\ \hline 8460.0 = 8460 \end{array}$$

25.
$$\begin{array}{r} 4.57 \\ \times\ \ 11.8 \\ \hline 3656 \\ 457\ \\ 457\ \ \\ \hline 53.926 \end{array}$$

27.
$$\begin{array}{r} 6523.7 \\ \times\ \ 0.001 \\ \hline 6.5237 \end{array}$$

29.
$$\begin{array}{r} 155.40 \\ \times\ \ \ \ 60 \\ \hline \$9324.00 = \$9324 \end{array}$$

31.
$$\begin{array}{r} 14.70 \\ \times\ \ \ \ 40 \\ \hline 588.00 = \$588 \end{array}$$

33.
$$\begin{array}{r} 19.2 \\ \times\ 15.5 \\ \hline 960 \\ 960\ \\ \hline 297.60 = 297.6 \text{ square feet} \end{array}$$

35.
$$\begin{array}{r} 36.90 \\ \times\ \ \ \ 18 \\ \hline 29520 \\ 3690\ \\ \hline \$664.20 \end{array}$$

37.　26.4
　　× 19.5
　　―――――
　　1320
　　2376
　　264
　　―――――
　514.80 = 514.8 miles

39. $2.86 \times 10 = 28.6$

41. $52.0 \times 100 = 5200$

43. $22.615 \times 1000 = 22,615$

45. $5.60982 \times 10,000 = 56,098.2$

47. $17,561.44 \times 10^2 = 1,756,144$

49. $816.32 \times 10^3 = 816,320$

51. $5.932 \times 1000 = 593.2$ centimeters

53. $2.98 \times 1000 = 2980$ meters

55.　　124.00　　　Amount left is
　　　　110.00
　　　　83.60　　　　820.00
　　　　76.00　　　− 572.00
　　　　44.60　　　――――――
　　　　44.60　　　　$284.00
　　　　44.60
　　　+　44.60
　　　――――――
　　　　$572.00

57.　254.2
　　× 19.6
　　――――
　15252
　22878
　2542
　――――
　4982.32 square yards

　　　　4982.32
　　　×　12.50
　　　――――――
　　　000000
　　　2491160
　　　996474
　　　498232
　　　――――――
　　　$62,279.00

59. To multiply by numbers such as 0.1, 0.01, 0.001, and 0.0001, count the number of decimal places in the first number. Then, in the other number, move the decimal point to the left from its present position the same number of decimal places as was in the first number.

Cumulative Review

61.　　86
　17)1462
　　136
　――――
　　102
　　102
　――――
　　　0

63.　　127
　48)6099
　　48
　――――
　129
　96
　――――
　339
　336
　――――
　　3
　127 R 3

65.　$0.73 billion
　−　$0.62 billion
　――――――――――
　$0.11 billion or $110,000,000

67.　2.510
　+ 1.963
　――――
　4.473
　$4.473 billion

How Am I Doing? Sections 3.1 - 3.4

1. Forty-seven and eight hundred thirteen thousandths

2. $\frac{567}{10,000} = 0.0567$

3. $2.11 = 2\frac{11}{100}$

4. $0.525 = \frac{525}{1000} = \frac{525 \div 25}{1000 \div 25} = \frac{21}{40}$

5. 1.59, 1.6, 1.601, 1.61

6. 123.49268 rounds to 123.5

7. 1.053458 rounds to 1.052

8. 17.98523 rounds to 17.99

9.
$$\begin{array}{r} 5.12 \\ 4.70 \\ 8.03 \\ +\ 1.60 \\ \hline 19.45 \end{array}$$

10.
$$\begin{array}{r} 24.613 \\ 0.273 \\ +\ 2.305 \\ \hline 27.191 \end{array}$$

11.
$$\begin{array}{r} 42.16 \\ -\ 31.57 \\ \hline 10.59 \end{array}$$

12.
$$\begin{array}{r} 26.000 \\ -\ 18.329 \\ \hline 7.671 \end{array}$$

13.
$$\begin{array}{r} 11.67 \\ \times\ 0.03 \\ \hline 0.3501 \end{array}$$

14. $4.7805 \times 1000 = 4780.5$

15. $0.0003796 \times 10^5 = 37.96$

16.
$$\begin{array}{r} 0.768 \\ \times\ 0.085 \\ \hline 3840 \\ 6144 \\ \hline 0.065280 \end{array}$$

17.
$$\begin{array}{r} 982 \\ \times\ 0.007 \\ \hline 6.874 \end{array}$$

18.
$$\begin{array}{r} 0.00052 \\ \times\ 0.006 \\ \hline 0.00000312 \end{array}$$

3.5 Exercises

1.
$$\begin{array}{r} 2.1 \\ 6\overline{)12.6} \\ \underline{12} \\ 6 \\ \underline{6} \\ 0 \end{array}$$

3.
$$\begin{array}{r} 17.83 \\ 4\overline{)71.32} \\ \underline{4} \\ 31 \\ \underline{28} \\ 33 \\ \underline{32} \\ 12 \\ \underline{12} \\ 0 \end{array}$$

5. $\begin{array}{r} 10.52 \\ 7\overline{)73.64} \\ \underline{7} \\ 36 \\ \underline{35} \\ 14 \\ \underline{14} \\ 0 \end{array}$

7. $\begin{array}{r} 136.5 \\ 0.6\overline{)81.90} \\ \underline{6} \\ 21 \\ \underline{18} \\ 39 \\ \underline{36} \\ 30 \\ \underline{30} \\ 0 \end{array}$

9. $\begin{array}{r} 9.05 \\ 0.04\overline{)0.3620} \\ \underline{36} \\ 20 \\ \underline{20} \\ 0 \end{array}$

11. $\begin{array}{r} 53.0 \\ 2.9\overline{)153.7} \\ \underline{145} \\ 87 \\ \underline{87} \\ 0 \end{array}$

13. $\begin{array}{r} 18 \\ 3.8\overline{)68.4} \\ \underline{38} \\ 304 \\ \underline{304} \\ 0 \end{array}$

15. $\begin{array}{r} 130 \\ 0.31\overline{)40.30} \\ \underline{31} \\ 93 \\ \underline{93} \\ 0 \end{array}$

17. $\begin{array}{r} 5.25 \\ 9\overline{)47.31} \\ \underline{45} \\ 23 \\ \underline{18} \\ 51 \\ \underline{45} \\ 6 \end{array}$

5.3

19. $\begin{array}{r} 1.24 \\ 1.9\overline{)2.360} \\ \underline{19} \\ 46 \\ \underline{38} \\ 80 \\ \underline{76} \\ 4 \end{array}$

1.2

21. $\begin{array}{r} 49.29 \\ 0.85\overline{)41.9010} \\ \underline{340} \\ 790 \\ \underline{765} \\ 251 \\ \underline{170} \\ 810 \\ \underline{765} \\ 45 \end{array}$

49.3

```
        94.206
    5)471.030
     45
     ──
     21
     20
     ──
     10
     10
     ──
     30
     30
     ──
      0
```

94.21

25.

```
          13.561
    1.8)24.4100
      18
      ──
      64
      54
      ──
     101
      90
      ──
     110
     108
     ───
      20
      18
      ──
       2
```

13.56

27.

```
        0.170
    36)6.125
     36
     ──
     252
     252
     ───
       5
```

0.17

29.

```
        0.0811
    7)0.5681
     56
     ──
      8
      7
      ─
     11
      7
      ─
      4
```

0.081

31.

```
           91.2643
    0.87)79.400000
       783
       ───
       110
        87
        ──
       230
       174
       ───
       560
       522
       ───
       380
       348
       ───
       320
       261
       ───
        59
```

91.264

33.

```
        123.2
    19)23410
     19
     ──
     44
     38
     ──
     61
     57
     ──
     40
     38
     ──
      2
```

123

35. $\begin{array}{r} 213.2 \\ 0.0046\overline{)0.98100} \end{array}$

$$\begin{array}{r} \underline{92} \\ 61 \\ \underline{46} \\ 150 \\ \underline{138} \\ 120 \\ \underline{92} \\ 28 \end{array}$$

213

37. $\begin{array}{r} 82.73 \\ 12\overline{)992.76} \end{array}$

$$\begin{array}{r} \underline{96} \\ 32 \\ \underline{24} \\ 87 \\ \underline{84} \\ 36 \\ \underline{36} \\ 0 \end{array}$$

\$82.73 per month

39. $\begin{array}{r} 27.27 \\ 13.2\overline{)360.000} \end{array}$

$$\begin{array}{r} \underline{264} \\ 960 \\ \underline{924} \\ 360 \\ \underline{264} \\ 960 \\ \underline{924} \\ 36 \end{array}$$

≈ 27.3 miles per gallon

41. $\begin{array}{r} 24 \\ 12.50\overline{)300.00} \end{array}$

$$\begin{array}{r} \underline{2500} \\ 5000 \\ \underline{5000} \\ 0 \end{array}$$

24 bouquets

43. $\begin{array}{r} 182 \\ 10.25\overline{)1865.50} \end{array}$

$$\begin{array}{r} \underline{1025} \\ 8405 \\ \underline{8200} \\ 2050 \\ \underline{2050} \\ 0 \end{array}$$

182 guests

45. $\begin{array}{r} 23.0 \\ 3.8\overline{)87.40} \end{array}$

$$\begin{array}{r} \underline{76} \\ 114 \\ \underline{114} \\ 0 \end{array}$$

The box contains 23 snowboards. The error was in putting in 1 less snowboard in the box than was required.

47. $0.8 \times n = 5.768$

$\begin{array}{r} 7.21 \\ 0.8\overline{)5.768} \end{array}$

$$\begin{array}{r} \underline{56} \\ 16 \\ \underline{16} \\ 8 \\ \underline{8} \\ 0 \end{array}$$

$n = 7.21$

49. $1.3 \times n = 1267.5$

$$\begin{array}{r} 975.0 \\ 1.3\overline{)1267.5} \\ \underline{117} \\ 97 \\ \underline{91} \\ 65 \\ \underline{65} \\ 0 \end{array}$$

$n = 975$

51. $n \times 0.098 = 4.312$

$$\begin{array}{r} 44.0 \\ 0.098\overline{)4.312} \\ \underline{392} \\ 392 \\ \underline{392} \\ 0 \end{array}$$

$n = 44$

53. $\frac{2.9356}{0.0716} \times \frac{10,000}{10,000} = \frac{29,356}{716}$

$$\begin{array}{r} 41 \\ 716\overline{)29356} \\ \underline{2864} \\ 716 \\ \underline{716} \\ 0 \end{array}$$

55. $2\frac{13}{16} - 1\frac{7}{8} = 2\frac{13}{16} - 1\frac{14}{16}$

$= 1\frac{29}{16} - 1\frac{14}{16}$

$= \frac{15}{16}$

57. $4\frac{1}{3} \div 2\frac{3}{5} = \frac{13}{3} \div \frac{13}{5}$

$= \frac{13}{3} \times \frac{5}{13}$

$= \frac{5}{3} = 1\frac{2}{3}$

59.

$$\begin{array}{r} 946 \\ -\ 475 \\ \hline 471 \end{array}$$

471 radios

61.

$$\begin{array}{r} 1443 \\ 946 \\ 880 \\ +\ 475 \\ \hline 3744 \end{array}$$

$\text{Average} = \frac{3744}{4}$

$= 936$ radios

3.6 Exercises

1. same quantity

3. The digits 8942 repeat.

5.

$$\begin{array}{r} 0.25 \\ 4\overline{)1.00} \\ \underline{8} \\ 20 \\ \underline{20} \\ 0 \end{array}$$

$\frac{1}{4} = 0.25$

7.

$$\begin{array}{r} 0.8 \\ 5\overline{)4.0} \\ \underline{40} \\ 0 \end{array}$$

$\frac{4}{5} = 0.8$

9. $$\begin{array}{r} 0.125 \\ 8\overline{)1.000} \\ \underline{8} \\ 20 \\ \underline{16} \\ 40 \\ \underline{40} \\ 0 \end{array}$$

$$\frac{1}{8} = 0.125$$

11. $$\begin{array}{r} 0.35 \\ 20\overline{)7.00} \\ \underline{60} \\ 100 \\ \underline{100} \\ 0 \end{array}$$

$$\frac{7}{20} = 0.35$$

13. $$\begin{array}{r} 0.62 \\ 50\overline{)31.00} \\ \underline{300} \\ 100 \\ \underline{100} \\ 0 \end{array}$$

$$\frac{31}{50} = 0.62$$

15. $$\begin{array}{r} 2.25 \\ 4\overline{)9.00} \\ \underline{8} \\ 10 \\ \underline{8} \\ 20 \\ \underline{20} \\ 0 \end{array}$$

$$\frac{9}{4} = 2.25$$

17. $$\begin{array}{r} 0.125 \\ 8\overline{)1.000} \\ \underline{8} \\ 20 \\ \underline{16} \\ 40 \\ \underline{40} \\ 0 \end{array}$$

$$2\frac{1}{8} = 2.125$$

19. $$\begin{array}{r} 0.4375 \\ 16\overline{)7.0000} \\ \underline{64} \\ 60 \\ \underline{48} \\ 120 \\ \underline{112} \\ 80 \\ \underline{80} \\ 0 \end{array}$$

$$1\frac{7}{16} = 1.4375$$

21. $$\begin{array}{r} 0.666 \\ 3\overline{)2.000} \\ \underline{18} \\ 20 \\ \underline{18} \\ 20 \\ \underline{18} \\ 2 \end{array}$$

$$\frac{2}{3} = 0.\overline{6}$$

23.
$$\begin{array}{r} 0.454 \\ 11\overline{)5.000} \\ \underline{44} \\ 60 \\ \underline{55} \\ 50 \\ \underline{44} \\ 6 \end{array}$$
$\frac{5}{11} = 0.\overline{45}$

25.
$$\begin{array}{r} 0.5833 \\ 12\overline{)7.0000} \\ \underline{60} \\ 100 \\ \underline{96} \\ 40 \\ \underline{36} \\ 40 \\ \underline{36} \\ 4 \end{array}$$
$3\frac{7}{12} = 3.58\overline{3}$

27.
$$\begin{array}{r} 0.277 \\ 18\overline{)5.000} \\ \underline{36} \\ 140 \\ \underline{126} \\ 140 \\ \underline{126} \\ 14 \end{array}$$
$2\frac{5}{18} = 2.2\overline{7}$

29.
$$\begin{array}{r} 0.3076 \\ 13\overline{)4.0000} \\ \underline{39} \\ 100 \\ \underline{91} \\ 90 \\ \underline{78} \\ 12 \end{array}$$
$\frac{4}{13} = 0.308$

31.
$$\begin{array}{r} 0.9047 \\ 21\overline{)19.0000} \\ \underline{189} \\ 10 \\ \underline{0} \\ 100 \\ \underline{84} \\ 160 \\ \underline{147} \\ 13 \end{array}$$
$\frac{19}{21}$ rounds to 0.905

33.
$$\begin{array}{r} 0.1458 \\ 48\overline{)7.0000} \\ \underline{48} \\ 220 \\ \underline{192} \\ 280 \\ \underline{240} \\ 400 \\ \underline{384} \\ 16 \end{array}$$
$\frac{7}{48}$ rounds to 0.146.

35.
$$\begin{array}{r} 1.2962 \\ 27\overline{)35.0000} \\ \underline{27} \\ 80 \\ \underline{54} \\ 260 \\ \underline{243} \\ 170 \\ \underline{162} \\ 80 \\ \underline{54} \\ 26 \end{array}$$

$\frac{35}{27}$ rounds to 1.296.

37.
$$\begin{array}{r} 0.4038 \\ 52\overline{)21.0000} \\ \underline{208} \\ 20 \\ \underline{00} \\ 200 \\ \underline{156} \\ 440 \\ \underline{416} \\ 24 \end{array}$$

$\frac{21}{52}$ rounds to 0.404.

39.
$$\begin{array}{r} 0.944 \\ 18\overline{)17.0} \\ \underline{162} \\ 80 \\ \underline{72} \\ 80 \\ \underline{72} \\ 8 \end{array}$$

$\frac{17}{18}$ rounds to 0.944.

41.
$$\begin{array}{r} 3.1428 \\ 7\overline{)22.0000} \\ \underline{21} \\ 10 \\ \underline{7} \\ 30 \\ \underline{28} \\ 20 \\ \underline{14} \\ 60 \\ \underline{56} \\ 4 \end{array}$$

$\frac{22}{7}$ rounds to 3.143.

43.
$$\begin{array}{r} 0.4736 \\ 19\overline{)9.0000} \\ \underline{76} \\ 140 \\ \underline{133} \\ 70 \\ \underline{57} \\ 130 \\ \underline{144} \\ 16 \end{array}$$

$3\frac{9}{19} = 3.474$

45.
$$\begin{array}{r} 0.875 \\ 8\overline{)7.000} \\ \underline{64} \\ 60 \\ \underline{56} \\ 40 \\ \underline{40} \\ 0 \end{array}$$

$\frac{7}{8} = 0.875 < 0.88$

47. $16\overline{)9.0000}$ quotient 0.5625

$$\begin{array}{r} 0.5625 \\ 16\overline{)9.0000} \\ \underline{80} \\ 100 \\ \underline{96} \\ 40 \\ \underline{32} \\ 80 \end{array}$$

$\frac{9}{16} = 0.5625$

$0.573 > \frac{9}{16}$

49.

$$\begin{array}{r} 0.125 \\ 8\overline{)1.000} \\ \underline{8} \\ 20 \\ \underline{16} \\ 40 \\ \underline{40} \\ 0 \end{array}$$

$\frac{1}{8} = 0.125$ inch

51.

$$\begin{array}{r} 0.375 \\ 8\overline{)3.000} \\ \underline{24} \\ 60 \\ \underline{56} \\ 40 \\ \underline{40} \\ 0 \end{array} \qquad \begin{array}{r} 0.500 \\ \underline{-\,0.375} \\ 0.125 \end{array}$$

It is too small by 0.125 inch.

53.

$$\begin{array}{r} 0.375 \\ 8\overline{)3.000} \\ \underline{24} \\ 60 \\ \underline{56} \\ 40 \\ \underline{40} \\ 0 \end{array} \qquad \begin{array}{r} 2.400 \\ \underline{-\,2.375} \\ 0.025 \end{array}$$

Yes; it is 0.025 inches too wide.

$2\frac{3}{8} = 2.375$

55. $2.4 + (0.5)^2 - 0.35$
$= 2.4 + 0.25 - 0.35$
$= 2.65 - 0.35$
$= 2.30$ or 2.3

57. $2.3 \times 3.2 - 5 \times 0.8 = 7.36 - 4.00 = 3.36$

59. $12 \div 0.03 - 50(0.5 + 1.5)^3$
$= 12 \div 0.03 - 50 \times (2)^3$
$= 12 \div 0.03 - 50(8)$
$= 400 - 400$
$= 0$

61. $(1.1)^3 + 2.6 \div 0.13 + 0.083$
$= 1.331 + 20 + 0.083$
$= 21.414$

63. $(14.73 - 14.61)^2 \div (1.18 + 0.82)$
$= (0.12)^2 \div 2$
$= 0.0144 \div 2$
$= 0.0072$

65. $(0.5)^3 + (3 - 2.6) \times 0.5$
$= (0.5)^3 + 0.4 \times 0.5$
$= 0.125 + 0.20$
$= 0.325$

67. $(0.76+4.24)\div 0.25+8.6$
$=5.00\div 0.25+8.6$
$=20.0+8.6$
$=28.6$

69. $(1.6)^3+(2.4)^2+18.666\div 3.05+4.86$
$=4.096+5.76+6.12+4.86$
$=20.836$

71.

$$\begin{array}{r} 0.5869297 \\ 8921\overline{)5236.0000000} \\ \underline{44605} \\ 77550 \\ \underline{71368} \\ 61820 \\ \underline{53526} \\ 82940 \\ \underline{80289} \\ 26510 \\ \underline{17842} \\ 86680 \\ \underline{80289} \\ 63910 \\ \underline{62447} \\ 1463 \end{array}$$

$\frac{5236}{8921}=0.586930$

73. a.
$$\begin{array}{r} 0.16\overline{16} \\ -\ 0.00\overline{16} \\ \hline 0.16 \end{array}$$

b.
$$\begin{array}{r} 0.1616\overline{16} \\ -\ 0.01666\overline{6} \\ \hline 0.144\overline{949} \end{array}$$

c. (b) is repeating and (a) is a nonrepeating decimal.

Cumulative Review

75. $\frac{25}{2}=12\frac{1}{2}$

$$\begin{array}{r} 12\frac{1}{2} \\ -\ \ 6\frac{3}{4} \\ \hline \end{array} \quad \begin{array}{r} 12\frac{2}{4} \\ -\ \ 6\frac{3}{4} \\ \hline \end{array} \quad \begin{array}{r} 11\frac{6}{4} \\ -\ 6\frac{3}{4} \\ \hline \end{array}$$

$5\frac{3}{4}$ feet deep

3.7 Exercises

1.
$$\begin{array}{r} 200,000,000 \\ +500,000,000 \\ \hline 700,000,000 \end{array}$$

3.
$$\begin{array}{r} 60,000 \\ -\ 30,000 \\ \hline 30,000 \end{array}$$

5. $5000\times 0.5=2500$

7. $900\div 60=15$

9. $4,000,000,000\div 20,000,000=\$200$

11.
$$\begin{array}{r} 650 \\ \times\ 7.5 \\ \hline 3250 \\ 4550 \\ \hline 4875.0 \end{array}$$
4875.0 kroners

13.
$$\begin{array}{r} 48.3 \\ \times\ 56.9 \\ \hline 4347 \\ 2898 \\ 2415 \\ \hline 2748.27 \end{array}$$
2748.27 square feet

15. $$\begin{array}{r} 96 \\ 0.12\overline{)11.52} \\ \underline{108} \\ 72 \\ \underline{72} \\ 0 \end{array}$$

96 molds

17. $$\begin{array}{r} 11.68 \\ 10.42 \\ \underline{+\,12.67} \\ 34.77 \end{array}$$

$$\begin{array}{r} 11.59 \\ 3\overline{)34.77} \\ \underline{3} \\ 4 \\ \underline{3} \\ 17 \\ \underline{15} \\ 27 \\ \underline{27} \\ 0 \end{array}$$

11.59 meters of rainfall per year.

19. $$\begin{array}{r} 24.0 \\ 0.75\overline{)18.00} \\ \underline{150} \\ 300 \\ \underline{300} \\ 0 \end{array}$$

24 servings

21. $$\begin{array}{r} 43.9 \\ 11.3 \\ \underline{+63.4} \\ 118.6 \end{array}$$

$$\begin{array}{r} 118.6 \\ \times\ \underline{10.65} \\ 5930 \\ 7116 \\ \underline{11860} \\ 1263.090 \end{array}$$

$1263.09

23. $$\begin{array}{r} 6\times 8\times 8.50 = 408 \\ \underline{+1.5\times 8.50\times 8 = 102} \\ \$510 \end{array}$$

25. $1.5\times 6.99 = 10.485$

$1.25\times 4.59 = 5.7375$

$$\begin{array}{r} 10.49 \\ \underline{+\ \ 5.73} \\ \$16.22 \end{array}$$

27. $$\begin{array}{r} 288.65 \\ \underline{\times\quad 60} \\ 17,319 \end{array}$$

$$\begin{array}{r} 17,319 \\ \underline{-\ 11,500} \\ 5819 \end{array}$$

He will pay $17,319 over 5 years.
He will pay $5819 more than the loan.

29. $$\begin{array}{r} 1.151 \\ 7\overline{)8.060} \\ \underline{7} \\ 1\,0 \\ \underline{7} \\ 36 \\ \underline{35} \\ 10 \\ \underline{7} \\ 3 \end{array}$$

Yes, by 0.149 milligram per liter.

31. $$\begin{array}{r} 137 \\ 126.4\overline{)17316.8} \\ \underline{1264} \\ 4676 \\ \underline{3792} \\ 8848 \\ \underline{8848} \\ 0 \end{array}$$

137 minutes

33. $$\begin{array}{r} 89.3 \\ -\ 66.4 \\ \hline 22.9 \end{array}$$ quadrillion Btu

35. $(33.1+43.8+66.4)\div 3$
$= 143.3 \div 3$
≈ 47.8
Approximately 47.8 quadrillion Btu;
47,800,000,000,000,000 Btu

Cumulative Review

37. $\frac{4}{7}+\frac{1}{2}\times\frac{2}{3}=\frac{4}{7}+\frac{1}{3}=\frac{12}{21}+\frac{7}{21}=\frac{19}{21}$

39. $\frac{7}{25}\times\frac{15}{42}=\frac{7\times 3\times 5}{5\times 5\times 2\times 3\times 7}=\frac{1}{10}$

Putting Your Skills to Work

1. $$\begin{array}{r} 72{,}900{,}000{,}000 \\ \times\qquad 22.3 \\ \hline 218{,}700{,}000{,}000 \\ 1458000000000 \\ 14580000000000 \\ \hline 1{,}625{,}670{,}000{,}000 \end{array}$$ miles

2. $$\begin{array}{l} 29557636363 \\ 55\overline{)1{,}625{,}670{,}000{,}000} \\ \underline{110} \\ 525 \\ \underline{495} \\ 306 \\ \underline{275} \\ 317 \\ \underline{275} \\ 420 \\ \underline{385} \\ 350 \\ \underline{330} \\ 200 \\ \underline{165} \\ 350 \\ \underline{330} \\ 200 \\ \underline{165} \\ 350 \end{array}$$

29,558,000,000 gallons

3. $\frac{1}{2}(72{,}900{,}000{,}000)+\frac{1}{2}(29{,}558{,}000{,}000)$
$= 36{,}450{,}000{,}000 + 14{,}779{,}000{,}000$
$= 51{,}229{,}000{,}000$ gallons

4. SUV: $\frac{20{,}000 \text{ mi}}{16 \text{ mi/gal}} = 1250 \text{ gal}$

Hybrid: $\frac{20{,}000 \text{ mi}}{55 \text{ mi/gal}} = 363.64 \text{ gal}$

Difference: $1250 - 363.64 = 886.36$ gal

Savings $= 886.36 \text{ gal} \times \frac{\$1.95}{\text{gal}} = \$1728$

Chapter 3 Review Problems

1. 13.672 = thirteen and six hundred seventy-two thousandths

2. Eighty-four hundred-thousandths

3. $\frac{7}{10} = 0.7$

4. $\frac{81}{100} = 0.81$

5. $1\frac{523}{1000} = 1.523$

6. $\frac{79}{10,000} = 0.0079$

7. $0.17 = \frac{17}{100}$

8. $\frac{0.036}{1} \times \frac{1000}{1000} = \frac{36}{1000} = \frac{9}{250}$

9. $34.24 = 34\frac{24}{100}$

$= 34\frac{6}{25}$

10. $1.00025 = 1\frac{25}{100,000} = 1\frac{1}{4000}$

11. $100\overline{)9.00}$ quotient .09

$\underline{900}$

0

$2\frac{9}{100} = 2.09$

12. $0.716 > 0.706$

13. $100\overline{)65.00}$ quotient .65

$\underline{600}$

500

$\underline{500}$

0

$\frac{65}{100} < 0.655$

14. $0.824 > 0.804$

15. 0.981, 0.918, 0.98, 0.901
0.981, 0.918, 0.980, 0.901
0.901, 0.918, 0.980, 0.981,
0.901, 0.918, 0.98, 0.981

16. 5.62, 5.2, 5.6, 5.26, 5.59
5.62, 5.20, 5.60, 5.26, 5.59
5.20, 5.26, 5.59, 5.60, 5.62
5.2, 5.26, 5.59, 5.6, 5.62

17. 0.704, 0.7045, 0.745, 0.754

18. 8.2, 8.27, 8.702, 8.72

19. 0.6$\underline{1}$3 rounds to 0.6

20. 19.20$\underline{7}$6 rounds to 19.21

21. 9.852$\underline{1}$5 rounds to 9.8522

22. \$156.$\underline{4}$8 rounds to \$156

23.
$$\begin{array}{r} 9.6 \\ 11.5 \\ 21.8 \\ +\ 34.7 \\ \hline 77.6 \end{array}$$

24.
$$\begin{array}{r} 1.800 \\ 2.603 \\ 0.520 \\ +\ 1.716 \\ \hline 6.639 \end{array}$$

25. $\begin{array}{r} 5.190 \\ -1.296 \\ \hline 3.894 \end{array}$

26. $\begin{array}{r} 182.422 \\ -\ 68.550 \\ \hline 113.872 \end{array}$

27. $\begin{array}{r} 0.098 \\ \times\ 0.032 \\ \hline 0196 \\ 0294 \\ \hline 0.003136 \end{array}$

28. $\begin{array}{r} 126.83 \\ \times\quad 7 \\ \hline 887.81 \end{array}$

29. $\begin{array}{r} 7.8 \\ \times\ 5.2 \\ \hline 156 \\ 390 \\ \hline 405.6 \end{array}$

30. $\begin{array}{r} 7053 \\ \times\ 0.34 \\ \hline 28212 \\ 21159 \\ \hline 2398.02 \end{array}$

31. $0.000613 \times 10^3 = 0.613$

32. $1.2354 \times 10^5 = 123{,}540$

33. $\begin{array}{r} 3.49 \\ \times\ 2.5 \\ \hline 1745 \\ 698 \\ \hline 8.725 \end{array}$

$8.73

34. $\begin{array}{r} 0.00258 \\ 0.07\overline{)0.0001806} \\ \underline{14} \\ 40 \\ \underline{35} \\ 56 \\ \underline{56} \\ 0 \end{array}$

35. $\begin{array}{r} 36.8 \\ 5.2\overline{)191.36} \\ \underline{156} \\ 353 \\ \underline{312} \\ 416 \\ \underline{416} \\ 0 \end{array}$

36. $\begin{array}{r} 232.9 \\ 8\overline{)1863.2} \\ \underline{16} \\ 26 \\ \underline{24} \\ 23 \\ \underline{16} \\ 72 \\ \underline{72} \\ 0 \end{array}$

37. $\begin{array}{r} 574.42 \\ 1.3\overline{)746.750} \end{array}$

$\underline{65}$
96
$\underline{91}$
57
$\underline{52}$
55
$\underline{52}$
30
$\underline{26}$
4

574.42 rounds to 574.4

38. $\begin{array}{r} 0.0589 \\ 0.06\overline{)0.003539} \end{array}$

$\underline{30}$
53
$\underline{48}$
59
$\underline{54}$
5

0.0589 rounds to 0.059

39. $\begin{array}{r} 0.277 \\ 18\overline{)5.000} \end{array}$

$\underline{36}$
140
$\underline{126}$
140
$\underline{126}$
14

$\frac{5}{18} = 0.2\overline{7}$

40. $\begin{array}{r} 0.175 \\ 40\overline{)7.000} \end{array}$

$\underline{40}$
300
$\underline{280}$
200
$\underline{200}$
0

$\frac{7}{40} = 0.175$

41. $\begin{array}{r} 0.833 \\ 6\overline{)5.000} \end{array}$

$\underline{48}$
20
$\underline{18}$
20
$\underline{18}$
2

$1\frac{5}{6} = 1.8\overline{3}$

42. $\begin{array}{r} 1.1875 \\ 16\overline{)19.0000} \end{array}$

$\underline{16}$
30
$\underline{16}$
140
$\underline{128}$
120
$\underline{112}$
80
$\underline{80}$
0

$\frac{19}{16} = 1.1875$

43.
$$\begin{array}{r} 0.7857 \\ 14\overline{)11.0000} \\ \underline{98} \\ 120 \\ \underline{112} \\ 80 \\ \underline{70} \\ 100 \\ \underline{98} \\ 2 \end{array}$$

$\frac{11}{14}$ rounds to 0.786

44.
$$\begin{array}{r} 0.3448 \\ 29\overline{)10.0000} \\ \underline{87} \\ 130 \\ \underline{116} \\ 140 \\ \underline{116} \\ 240 \\ \underline{232} \\ 8 \end{array}$$

0.345

45.
$$\begin{array}{r} 0.2941 \\ 17\overline{)5.0000} \\ \underline{34} \\ 160 \\ \underline{153} \\ 70 \\ \underline{68} \\ 20 \\ \underline{17} \\ 3 \end{array}$$

$2\frac{5}{17}$ rounds to 2.294

46.
$$\begin{array}{r} .3913 \\ 23\overline{)9.0000} \\ \underline{69} \\ 210 \\ \underline{207} \\ 30 \\ \underline{23} \\ 70 \\ \underline{69} \\ 1 \end{array}$$

$3\frac{9}{23} = 3.391$

47. $2.3 \times 1.82 + 3 \times 5.12$
$= 4.186 + 15.36$
$= 19.546$

48. $0.03 + (1.2)^2 - 5.3 \times 0.06$
$= 0.03 + 1.44 - 5.3 \times 0.06$
$= 0.03 + 1.44 - 0.318$
$= 1.47 - 0.318$
$= 1.152$

49. $(1.02)^3 + 5.76 \div 1.2 \times 0.05$
$= 1.061208 + 5.76 \div 1.2 \times 0.05$
$= 1.061208 + 4.8 \times 0.05$
$= 1.061208 + 0.24$
$= 1.301208$

50. $2.4 \div (2 - 1.6)^2 + 8.13$
$= 2.4 \div (0.4)^2 + 8.13$
$= 2.4 \div 0.16 + 8.13$
$= 15 + 8.13$
$= 23.13$

51.
$$\begin{array}{r} 2398.26 \\ -\ 1959.07 \\ \hline 439.19 \end{array}$$

52. $32.15 \times 0.02 \times 10^2$

$= 32.15 \times 0.02 \times 100$

$= 0.643 \times 100$

$= 64.3$

53. $1.809 - 0.62 + 3.27$

$= 1.189 + 3.27$

$= 4.459$

54.
$$\begin{array}{r} .904 \\ 2.3\overline{)2.0792} \\ \underline{2.07} \\ 92 \\ \underline{92} \\ 0 \end{array}$$

0.904

55. $8 \div 0.4 + 0.1 \times (0.2)^2$

$= 20 + 0.1 \times 0.04$

$= 20 + 0.004$

$= 20.004$

56. $(3.8 - 2.8)^3 \div (0.5 + 0.3)$

$= 1^3 \div 0.8$

$= 1 \div 0.8$

$= 1.25$

57. $\text{Tickets} = 228 + 2.5 \times 388 + 3 \times 430$

$= 228 + 970 + 1290$

$= 2488$

$\text{Not tickets} = 2600 - 2488$

$= 112 \text{ people}$

58.
$$\begin{array}{r} 26325.8 \\ -\ 26005.8 \\ \hline 320.0 \end{array}$$

$$\begin{array}{r} 24.80 \\ 12.9\overline{)320.000} \\ \underline{258} \\ 620 \\ \underline{516} \\ 1040 \\ \underline{1032} \\ 80 \\ \underline{0} \\ 80 \end{array}$$

24.8 miles per gallon

59.
$$\begin{array}{r} 189.60 \\ \times\ \ \ 48 \\ \hline 151680 \\ \underline{75840} \\ \$9100.80 \end{array}$$

$$\begin{array}{r} 9100.80 \\ -\ 6930.50 \\ \hline \$2170.30 \text{ extra} \end{array}$$

60.
$$\begin{array}{r} 8.26 \\ \times\ \ 38 \\ \hline 6608 \\ \underline{2478} \\ \$313.88 \end{array}$$

He will earn more at the ABC Company.

61.
$$\begin{array}{r} 0.0025 \\ 12\overline{)0.0300} \\ \underline{24} \\ 60 \\ \underline{60} \\ 0 \end{array}$$

$$\begin{array}{r} 0.0025 \\ \underline{-0.0020} \\ 0.0005 \end{array}$$

No; by 0.0005 milligram per liter.

62.
$$\begin{array}{r} 15.748 \\ 2.54\overline{)40.00000} \\ \underline{254} \\ 1460 \\ \underline{1270} \\ 1900 \\ \underline{1778} \\ 1220 \\ \underline{1016} \\ 2040 \\ \underline{2032} \\ 8 \end{array}$$

15.75 inches

63. a. $\text{Fence} = 2\times 18.3 + 2\times 9.6$
$= 36.6 + 19.2$
$= 55.8$ feet

b.
$$\begin{array}{r} 18.3 \\ \underline{\times 9.6} \\ 1098 \\ \underline{1647} \end{array}$$

175.68 square feet

64.
$$\begin{array}{r} 75.5 \\ \underline{\times 18.5} \\ 3775 \\ 6040 \\ \underline{755} \\ 1396.75 \end{array}$$

1396.75 square feet

65. Galeton to Wellsboro
$5.7 + 18.4 = 24.1$ miles
Coudersport to Gaines
$16.3 + 8.2 + 5.7 = 30.2$ miles
Difference
$30.2 - 24.1 = 6.1$ miles

66.
$$\begin{array}{r} 118.9 \\ 25.6 \\ 18.9 \\ 43.9 \\ 22.6 \\ 13.8 \\ \underline{+\ 16.2} \\ 259.9 \end{array}$$

259.9 feet around the field

67.
$$\begin{array}{r} 212.50 \\ \underline{\times \quad 60} \\ \$12,750.00 \end{array}$$

$$\begin{array}{r} 199.50 \\ \underline{\times \quad 60} \\ 11,970.00 \\ \underline{+ \quad 285.00} \\ \$12,255.00 \end{array}$$

They should change to the new loan.

68.
$$\begin{array}{r} 603 \\ \underline{-341} \\ 262 \end{array}$$

Increase is $262

69.
$$\begin{array}{r} 810 \\ -603 \\ \hline \$207 \end{array}$$
Average is \$207

70. $479 \div 30 = 15.97$
Average daily benefit is \$15.97.

71. $\frac{720}{30} = 24$
Average is \$24.00.

72.
$$\begin{array}{r} 810 \\ -479 \\ \hline 331 \end{array} \quad \begin{array}{r} 810 \\ +331 \\ \hline 1141 \end{array}$$
$$\begin{array}{r} 38.033 \\ 30\overline{)1141.0} \\ 90 \\ \hline 241 \\ 240 \\ \hline 100 \\ 90 \\ \hline 100 \\ 90 \\ \hline 10 \end{array}$$
\$38.03

73.
$$\begin{array}{r} 810 \\ +207 \\ \hline 1017 \end{array}$$
Then $\frac{1017}{30} = 33.90$
Average is \$33.90.

How Am I Doing? Chapter 3 Test

1. Twelve and forty-three thousandths

2. $\frac{3977}{10,000} = 0.3977$

3. $7.15 = 7\frac{15}{100} = 7\frac{3}{20}$

4. $0.261 = \frac{261}{1000}$

5. 2.19, 2.91, 2.9, 2.907
2.190, 2.910, 2.900, 2.907

2.190, 2.91, 2.9, 2.907, 2.910
2.19, 2.9, 2.907, 2.91

6. 78.65$\underline{6}$2 rounds to 78.66

7. 0.341$\underline{7}$52 rounds to 0.0342

8.
$$\begin{array}{r} 96.200 \\ 1.348 \\ +\ 2.150 \\ \hline 99.698 \end{array}$$

9.
$$\begin{array}{r} 17.00 \\ 2.10 \\ 16.80 \\ 0.04 \\ +\ 1.59 \\ \hline 37.53 \end{array}$$

10.
$$\begin{array}{r} 1.0075 \\ -\ 0.9096 \\ \hline 0.0979 \end{array}$$

11.
$$\begin{array}{r} 72.300 \\ -\ 1.145 \\ \hline 71.155 \end{array}$$

12.
$$\begin{array}{r} 8.31 \\ \times\ 0.07 \\ \hline 0.5817 \end{array}$$

13. $2.189 \times 10^3 = 2189$

14. $\begin{array}{r} 0.1285 \\ 0.08\overline{)0.010280} \\ \underline{8} \\ 22 \\ \underline{16} \\ 68 \\ \underline{64} \\ 40 \\ \underline{40} \\ 0 \end{array}$

15. $\begin{array}{r} 47 \\ 0.69\overline{)32.43} \\ \underline{276} \\ 483 \\ \underline{483} \\ 0 \end{array}$

16. $\begin{array}{r} 1.2 \\ 9\overline{)11.0} \\ \underline{9} \\ 20 \\ \underline{18} \\ 2 \end{array}$

$\frac{11}{9} = 1.\overline{2}$

17. $\begin{array}{r} 0.875 \\ 8\overline{)7.000} \\ \underline{64} \\ 60 \\ \underline{56} \\ 40 \\ \underline{40} \\ 0 \end{array}$

$\frac{7}{8} = 0.875$

18. $(0.3)^3 + 1.02 \div 0.5 - 0.58$
$= 0.027 + 1.02 \div 0.5 - 0.58$
$= 0.027 + 2.04 - 0.58$
$= 2.067 - 0.58$
$= 1.487$

19. $19.36 \div (0.24 + 0.26) \times (0.4)^2$
$= 19.36 \div 0.5 \times 0.16$
$= 38.72 \times 0.16$
$= 6.1952$

20. $\begin{array}{r} 1.41 \\ \times\ 8.5 \\ \hline 705 \\ 1128 \\ \hline 11.985 \end{array}$

\$11.99

21. $\begin{array}{r} 42780.5 \\ -\ 42620.5 \\ \hline 160.0 \end{array}$

$\begin{array}{r} 18.82 \\ 8.5\overline{)160.000} \\ \underline{85} \\ 750 \\ \underline{680} \\ 700 \\ \underline{680} \\ 200 \\ \underline{170} \\ 30 \end{array}$

18.8 miles per gallon

22.
$$\begin{array}{r} 8.01 \\ 5.03 \\ +\ 8.53 \\ \hline 21.57 \end{array}$$

$$\begin{array}{r} 25.00 \\ -\ 21.57 \\ \hline 3.43 \end{array}$$

3.43 centimeters less

23. $\text{Time} = 40 + 1.5 \times 9$
$= 40 + 13.5$
$= 53.5$ hours
$\text{Salary} = \$7.30 \times 53.5$
$= \$390.55$

Cumulative Test for Chapters 1 - 3

1. 38,056,954 = Thirty-eight million, fifty-six thousand, nine hundred fifty-four

2.
$$\begin{array}{r} 156,028 \\ 301,579 \\ +\quad 21,980 \\ \hline 479,587 \end{array}$$

3.
$$\begin{array}{r} 1,091,000 \\ -1,036,520 \\ \hline 54,480 \end{array}$$

4.
$$\begin{array}{r} 589 \\ \times\quad 67 \\ \hline 4123 \\ 3534 \\ \hline 39,463 \end{array}$$

5.
$$\begin{array}{r} 258 \\ 17\overline{)4386} \\ \underline{34} \\ 98 \\ \underline{85} \\ 136 \\ \underline{136} \\ 0 \end{array}$$

6. $20 \div 4 + 2^5 - 7 \times 3$
$= 20 \div 4 + 32 - 7 \times 3$
$= 5 + 32 - 21$
$= 37 - 21$
$= 16$

7. $\frac{18}{45} = \frac{18 \div 9}{45 \div 9} = \frac{2}{5}$

8. $4\frac{1}{3} + 3\frac{1}{6} = 4\frac{2}{6} + 3\frac{1}{6}$
$= 7\frac{3}{6}$
$= 7\frac{1}{2}$

9. $\frac{23}{35} - \frac{2}{5} = \frac{23}{35} - \frac{2}{5} \times \frac{7}{7}$
$= \frac{23}{35} - \frac{14}{35}$
$= \frac{9}{35}$

10. $\frac{7}{10} \times \frac{5}{3} - \frac{5}{12} \times \frac{1}{2} = \frac{7}{6} - \frac{5}{24}$
$= \frac{7}{6} \times \frac{4}{4} - \frac{5}{24}$
$= \frac{28}{24} - \frac{5}{24}$
$= \frac{23}{24}$

11. $52 \div 3\frac{1}{4} = 52 \div \frac{13}{4}$
$= 52 \times \frac{4}{13}$
$= 16$

12. $$1\frac{3}{8}\div\frac{5}{12}=\frac{11}{8}\div\frac{5}{12}$$
$$=\frac{11}{8}\times\frac{12}{5}$$
$$=\frac{11\times4\times3}{4\times2\times5}$$
$$=\frac{33}{10}=3\frac{3}{10}$$

13. $60,000\times400,000=24,000,000,000$

14. $\frac{39}{1000}=0.039$

15. 2.01, 2.1, 2.11, 2.12, 20.1

16. $26.079\underline{8}4$ rounds to 26.080

17.
$$\begin{array}{r} 1.90 \\ 2.36 \\ 15.20 \\ +\ 0.08 \\ \hline 19.54 \end{array}$$

18.
$$\begin{array}{r} 28.100 \\ -\ 14.982 \\ \hline 13.118 \end{array}$$

19.
$$\begin{array}{r} 56.8 \\ \times\ 0.02 \\ \hline 1.136 \end{array}$$

20. $0.1823\times1000=182.3$

21.
$$\begin{array}{r} 1.058 \\ 0.06\overline{)0.06348} \\ \underline{6} \\ 3 \\ \underline{0} \\ 34 \\ \underline{30} \\ 48 \\ \underline{48} \\ 0 \end{array}$$

22.
$$\begin{array}{r} 0.8125 \\ 16\overline{)13.0000} \\ \underline{128} \\ 20 \\ \underline{16} \\ 40 \\ \underline{32} \\ 80 \\ \underline{80} \\ 0 \end{array}$$

$\frac{13}{16}=0.8125$

23. $$1.44\div0.12+(0.3)^3+1.56$$
$$=1.44\div0.12+0.027+1.57$$
$$=12+0.027+1.57$$
$$=12.027+1.57$$
$$=13.597$$

24. a.
$$\begin{array}{r} 10.5 \\ \times\ 10.5 \\ \hline 525 \\ \underline{1050} \\ 110.25 \end{array}$$
110.25 square feet

b.
$$\begin{array}{r} 10.5 \\ \times\ \ 4 \\ \hline 42.0 \end{array}$$
42.0 feet

25.

$$\begin{array}{r} 199.36 \\ 1.03 \\ 166.35 \\ +\quad 93.50 \\ \hline 460.24 \end{array}$$

$$\begin{array}{r} 90.00 \\ 37.49 \\ +137.18 \\ \hline 264.67 \end{array}$$

$$\begin{array}{r} 460.24 \\ -264.67 \\ \hline 195.57 \end{array}$$

\$195.57

26.

$$\begin{array}{r} 60 \\ 320.50\overline{)19,230.00} \\ \underline{19,230.0} \\ 0 \end{array}$$

60 months

Chapter 4

4.1 Exercises

1. ratio

3. 5 to 8

5. $6:18 = \frac{6}{18} = \frac{6 \div 6}{18 \div 6} = \frac{1}{3}$

7. $21:18 = \frac{21}{18} = \frac{21 \div 3}{18 \div 3} = \frac{7}{6}$

9. $150:225 = \frac{150}{225} = \frac{150 \div 75}{225 \div 75} = \frac{2}{3}$

11. $165 \text{ to } 90 = \frac{165}{90} = \frac{165 \div 15}{90 \div 15} = \frac{11}{6}$

13. $60 \text{ to } 64 = \frac{60}{64} = \frac{60 \div 4}{64 \div 4} = \frac{15}{16}$

15. $28 \text{ to } 42 = \frac{28}{42} = \frac{28 \div 14}{42 \div 14} = \frac{2}{3}$

17. $32 \text{ to } 20 = \frac{32}{20} = \frac{32 \div 4}{20 \div 4} = \frac{8}{5}$

19. $8 \text{ ounces to } 12 \text{ ounces} = \frac{8}{12} = \frac{8 \div 4}{12 \div 4} = \frac{2}{3}$

21. 39 kilograms to 26 kilograms

$= \frac{39}{26} = \frac{39 \div 13}{26 \div 13} = \frac{3}{2}$

23. $82 \text{ to } 160 = \frac{82}{160} = \frac{82 \div 2}{160 \div 2} = \frac{41}{80}$

25. 312 yards to 24 yards

$= \frac{312}{24} = \frac{312 \div 24}{24 \div 24} = \frac{13}{1}$

27. $2\frac{1}{2}$ pounds to $4\frac{1}{4}$ pounds

$= \frac{2\frac{1}{2}}{4\frac{1}{4}} = 2\frac{1}{2} \div 4\frac{1}{4} = \frac{5}{2} \div \frac{17}{4}$

$= \frac{5}{2} \times \frac{4}{17} = \frac{10}{17}$

29. $\frac{165}{286} = \frac{165 \div 15}{285 \div 15} = \frac{11}{19}$

31. $\frac{35}{165} = \frac{35 \div 5}{165 \div 5} = \frac{7}{33}$

33. $\frac{205}{1225} = \frac{205 \div 5}{1225 \div 5} = \frac{41}{245}$

35. $\frac{450}{205} = \frac{450 \div 5}{205 \div 5} = \frac{90}{41}$

37. $\frac{44}{704} = \frac{44 \div 44}{704 \div 44} = \frac{1}{16}$

39. $\frac{\$40}{16 \text{ magazines}} = \frac{\$40 \div 8}{16 \text{ magazines} \div 8}$

$= \frac{\$5}{2 \text{ magazines}}$

41. $\frac{\$170}{12 \text{ bushes}} = \frac{\$170 \div 2}{12 \text{ bushes} \div 2} = \frac{\$85}{6 \text{ bushes}}$

43. $\frac{\$114}{12 \text{ CDs}} = \frac{\$114 \div 3}{12 \text{ CDs} \div 3} = \frac{\$38}{4 \text{ CDs}} = \frac{\$19}{2 \text{ CDs}}$

45. $\frac{6150 \text{ rev}}{15 \text{ miles}} = \frac{6150 \text{ rev} \div 15}{15 \text{ miles} \div 15}$

$= \frac{410 \text{ rev}}{1 \text{ mile}} = 410 \text{ rev/mile}$

47. $\frac{\$330,000}{12 \text{ employees}} = \frac{\$330,000 \div 12}{12 \text{ employees} \div 12}$
$= \frac{\$27,500}{1 \text{ employee}}$
$= \$27,500/\text{employee}$

49. $\frac{\$520}{40 \text{ hours}} = \$13/\text{hour}$

51. $\frac{192 \text{ miles}}{12 \text{ gallons}} = 16 \text{ mi/gal}$

53. $\frac{1800 \text{ people}}{20 \text{ sq. mi.}} = 90 \text{ people/sq. mi.}$

55. $\frac{2250 \text{ pencils}}{18 \text{ boxes}} = 125 \text{ pencils/box}$

57. $\frac{297 \text{ mi}}{4.5 \text{ hr}} = 66 \text{ mi/hr}$

59. $\frac{475 \text{ patients}}{25 \text{ doctors}} = 19 \text{ patients/doctor}$

61. $\frac{60 \text{ eggs}}{12 \text{ chickens}} = 5 \text{ eggs/chicken}$

63. $\frac{\$3870}{129 \text{ shares}} = \$30/\text{share}$

65. Profit $= 1200 - 760 = \$440$
$\frac{\$440}{80 \text{ calendars}} = \$5.50/\text{calendar}$

67. a. 16-ounce box: $\frac{\$1.28}{16 \text{ ounces}} = \$0.08/\text{oz}$
24-ounce box: $\frac{\$1.68}{24 \text{ ounces}} = \$0.07/\text{oz}$

b. $\begin{array}{r} 0.08 \\ -\ 0.07 \\ \hline \$0.01/\text{ounce} \end{array}$

c. $48(0.01) = \$0.48$

69. a. $\frac{3978 \text{ moose}}{306 \text{ acres}} = 13 \text{ moose/acre}$

b. $\frac{5520 \text{ moose}}{460 \text{ acres}} = 12 \text{ moose/acre}$

c. North Slope

71. a. $\frac{\$12,862.50}{525 \text{ shares}} = \$24.50/\text{share}$

b. $\frac{\$781}{355 \text{ shares}} = \$2.20/\text{share}$

c. $\begin{array}{r} 24.50 \\ -\ 2.20 \\ \hline \$22.30/\text{share} \end{array}$

73. Design:
$\frac{750 \text{ meters per second}}{330 \text{ meters per second}} = \text{Mach } 2.3$
Modify:
$\frac{810 \text{ meters per second}}{330 \text{ meters per second}} = \text{Mach } 2.5$
$\begin{array}{r} 2.5 \\ -2.3 \\ \hline 0.2 \end{array}$ Increased by Mach 0.2

Cumulative Review

75. $\begin{array}{r} 2\frac{1}{4} \\ +\ \frac{3}{8} \\ \hline \end{array}$ $\begin{array}{r} 2\frac{2}{8} \\ +\ \frac{3}{8} \\ \hline 2\frac{5}{8} \end{array}$

77. $\frac{3}{5} \times \frac{5}{8} - \frac{2}{3} \times \frac{1}{4} = \frac{3}{8} - \frac{2}{12}$
$= \frac{9}{24} - \frac{4}{24}$
$= \frac{5}{24}$

79. $12 \times 5.2 = 62.4$ sq yd
$\frac{\$764.40}{62.4 \text{ sq yd}} = \$12.25/\text{sq yd}$

4.2 Exercises

1. equal

3. $\frac{18}{9}=\frac{2}{1}$

5. $\frac{20}{36}=\frac{5}{9}$

7. $\frac{220}{11}=\frac{400}{20}$

9. $\frac{5\frac{1}{2}}{16}=\frac{7\frac{2}{3}}{23}$

11. $\frac{6.5}{14}=\frac{13}{28}$

13. $\frac{3\text{ inches}}{40\text{ miles}}=\frac{27\text{ inches}}{360\text{ miles}}$

15. $\frac{\$28}{6\text{ tables}}=\frac{\$98}{21\text{ tables}}$

17. $\frac{3\text{ hours}}{\$525}=\frac{7\text{ hours}}{\$1225}$

19. $\frac{3\text{ teaching assistants}}{40\text{ children}}=\frac{21\text{ teaching assistants}}{280\text{ children}}$

21. $\frac{4800\text{ people}}{3\text{ restaurants}}=\frac{11{,}200\text{ people}}{7\text{ restaurants}}$

23. $\frac{10}{25}\stackrel{?}{=}\frac{6}{15}$

$10\times 15\stackrel{?}{=}25\times 6$

$150=150$ True

It is a proportion.

25. $\frac{11}{7}\stackrel{?}{=}\frac{20}{13}$

$11\times 13\stackrel{?}{=}7\times 2$

$143\neq 140$

It is not a proportion.

27. $\frac{17}{75}\stackrel{?}{=}\frac{22}{100}$

$17\times 100\stackrel{?}{=}75\times 22$

$1700\neq 1650$

It is not a proportion.

29. $\frac{102}{120}\stackrel{?}{=}\frac{85}{100}$

$102\times 100\stackrel{?}{=}120\times 85$

$10,200=10,200$

It is a proportion.

31. $\frac{2.5}{4}\stackrel{?}{=}\frac{7.5}{12}$

$2.5\times 12\stackrel{?}{=}4\times 7.5$

$30=30$

It is a proportion.

33. $\frac{3}{17}\stackrel{?}{=}\frac{4.5}{24.5}$

$3\times 24.5\stackrel{?}{=}17\times 4.5$

$73.5\neq 76.5$

If is not a proportion.

35. $\frac{2}{4\frac{3}{4}}\stackrel{?}{=}\frac{8}{19}$

$2\times 19\stackrel{?}{=}4\frac{3}{4}\times 8$

$38\stackrel{?}{=}\frac{19}{4}\times 8$

$38=38$

It is a proportion.

37. $\frac{7\frac{1}{3}}{3}\stackrel{?}{=}\frac{23}{9}$

$7\frac{1}{3}\times 9\stackrel{?}{=}3\times 23$

$66\neq 69$

It is not a proportion.

39. $\frac{\frac{1}{4}}{2} \stackrel{?}{=} \frac{\frac{7}{20}}{2.8}$

$\frac{1}{2} \times 2.8 \stackrel{?}{=} 2 \times \frac{7}{20}$

$0.7 = 0.7$

It is a proportion.

41. $\frac{135 \text{ miles}}{3 \text{ hours}} \stackrel{?}{=} \frac{225 \text{ miles}}{5 \text{ hours}}$

$135 \times 5 \stackrel{?}{=} 3 \times 25$

$675 = 675$

It is a proportion.

43. $\frac{166 \text{ gallons}}{14 \text{ acres}} \stackrel{?}{=} \frac{249 \text{ gallons}}{21 \text{ acres}}$

$166 \times 21 \stackrel{?}{=} 14 \times 249$

$3486 = 3486$

It is a proportion.

45. $\frac{21 \text{ homeruns}}{96 \text{ games}} \stackrel{?}{=} \frac{18 \text{ homeruns}}{81 \text{ games}}$

$21 \times 81 \stackrel{?}{=} 96 \times 18$

$1701 \neq 1728$

It is not a proportion.

47. $\frac{22}{132} \stackrel{?}{=} \frac{32}{160}$

$22 \times 160 \stackrel{?}{=} 132 \times 32$

$3520 \neq 4224$ False

49. a. $\frac{550 \text{ miles}}{15 \text{ hours}} \stackrel{?}{=} \frac{230 \text{ miles}}{6 \text{ hours}}$

$550 \times 6 \stackrel{?}{=} 15 \times 230$

$3300 \neq 3450$ No

b. The bus is traveling at a faster rate.

51. $\frac{75 \text{ feet}}{20 \text{ feet}} \stackrel{?}{=} \frac{105 \text{ feet}}{28 \text{ feet}}$

$75 \times 28 \stackrel{?}{=} 20 \times 105$

$2100 = 2100$ Yes

53. a. $\frac{169}{221} = \frac{169 \div 13}{221 \div 13} = \frac{13}{17}$

$\frac{247}{323} = \frac{247 \div 19}{323 \div 19} = \frac{13}{17}$

True

b. $\frac{169}{221} \stackrel{?}{=} \frac{247}{323}$

$169 \times 323 \stackrel{?}{=} 221 \times 247$

$54,587 = 54,587$ True

c. For most students it is faster to multiply than to reduce fractions.

Cumulative Review

55.

$$\begin{array}{r} 5.92 \\ \times \;\; 3.04 \\ \hline 608 \\ 2736 \\ 1520 \\ \hline 17.9968 \end{array}$$

57.

$$\begin{array}{r} 25.8 \\ 7.03\overline{)181.374} \\ \underline{1406} \\ 4077 \\ \underline{3515} \\ 5624 \\ \underline{5624} \\ 0 \end{array}$$

How Am I Doing? Sections 4.1 - 4.2

1. $\frac{13}{18}$

2. $\frac{44}{220} = \frac{1}{5}$

3. $\frac{72}{16} = \frac{9}{2}$

4. $\frac{121}{132} = \frac{11}{12}$

5. a. $\frac{70}{240} = \frac{7}{24}$

b. $\frac{22}{240} = \frac{11}{120}$

6. $\frac{9}{300} = \frac{3 \text{ flight attendants}}{100 \text{ passengers}}$

7. $\frac{620}{840} = \frac{31 \text{ gallons}}{42 \text{ square feet}}$

8. $\frac{122}{4} = 30.5$ miles/hour

9. $\frac{435}{15} = \$29$ per CD player

10. $\frac{2400}{15} = 160$ cookies/pound of cookie dough

11. $\frac{13}{40} = \frac{39}{120}$

12. $\frac{116}{158} = \frac{29}{37}$

13. $\frac{33 \text{ nautical miles}}{2 \text{ hours}} = \frac{49.5 \text{ nautical miles}}{3 \text{ hours}}$

14. $\frac{3000 \text{ shoes}}{\$370} = \frac{7500 \text{ shoes}}{\$925}$

15. $\frac{14}{31} = \frac{42}{93}$

$14(93) = 31(42)$

$1302 = 1302$

It is a proportion.

16. $\frac{17}{33} = \frac{19}{45}$

$45(17) = 33(19)$

$765 \neq 627$

It is not a proportion.

17. $\frac{5.6}{3.2} = \frac{112}{64}$

$5.6(64) = 3.2(112)$

$358.4 = 358.4$

It is a proportion.

18. $\frac{2\frac{1}{2}}{3\frac{1}{3}} = \frac{35}{46}$

$\frac{\frac{5}{2}}{\frac{10}{3}} = \frac{35}{46}$

$\frac{5}{2}(46) = \frac{10}{3}(35)$

$115 \neq 116\frac{2}{3}$

It is not a proportion.

19. $\frac{670}{1541} = \frac{820}{1886}$

$670(1886) = 1541(820)$

$1,263,620 = 1,263,620$

It is a proportion.

20. $\frac{30}{4} = \frac{3000}{400}$

$30(400) = 4(3000)$

$12,000 = 12,000$

It is a proportion.

4.3 Exercises

1. Divide each side of the equation by the number a. Calculate $\frac{b}{a}$. The value of n is $\frac{b}{a}$.

3. $12\times n=132$

$$\frac{12\times n}{12}=\frac{132}{12}$$
$$n=11$$

5. $3\times n=16.8$

$$\frac{3\times n}{3}=\frac{16.8}{3}$$
$$n=5.6$$

7. $n\times 11.4=57$

$$\frac{n\times 11.4}{11.4}=\frac{57}{11.4}$$
$$n=5$$

9. $50.4=6.3\times n$

$$\frac{50.4}{6.3}=\frac{6.3\times n}{6.3}$$
$$8=n$$

11. $\frac{3}{4}\times n=26$

$$\frac{\frac{3}{4}\times n}{\frac{3}{4}}=\frac{26}{\frac{3}{4}}$$
$$n=26\div\frac{3}{4}$$
$$=26\times\frac{4}{3}$$
$$=\frac{104}{3}=34\tfrac{2}{3}$$

13. $\frac{n}{20}=\frac{3}{4}$

$$n\times 4=20\times 3$$
$$\frac{n\times 4}{4}=\frac{60}{4}$$
$$n=15$$

Check:

$$\frac{15}{20}\stackrel{?}{=}\frac{3}{4}$$
$$4\times 15\stackrel{?}{=}20\times 3$$
$$60=60$$

15. $\frac{6}{n}=\frac{3}{8}$

$$6\times 8=n\times 3$$
$$\frac{48}{3}=\frac{n\times 3}{3}$$
$$16=n$$

Check:

$$\frac{6}{16}\stackrel{?}{=}\frac{3}{8}$$
$$6\times 8\stackrel{?}{=}16\times 3$$
$$48=48$$

17. $\frac{12}{40}=\frac{n}{25}$

$$12\times 25=40\times n$$
$$\frac{300}{40}=\frac{40\times n}{40}$$
$$7.5=n$$

Check:

$$\frac{12}{40}\stackrel{?}{=}\frac{7.5}{25}$$
$$12\times 25\stackrel{?}{=}40\times 7.5$$
$$300=300$$

19. $\frac{50}{100} = \frac{2.5}{n}$

$50 \times n = 100 \times 2.5$

$\frac{50 \times n}{50} = \frac{250}{50}$

$n = 5$

Check:

$\frac{50}{100} \stackrel{?}{=} \frac{2.5}{5}$

$50 \times 5 \stackrel{?}{=} 100 \times 2.5$

$250 = 250$

21. $\frac{n}{6} = \frac{150}{12}$

$n \times 12 = 6 \times 150$

$\frac{n \times 12}{12} = \frac{900}{12}$

$n = 75$

Check:

$\frac{75}{6} \stackrel{?}{=} \frac{150}{12}$

$75 \times 12 \stackrel{?}{=} 6 \times 150$

$900 = 900$

23. $\frac{15}{4} = \frac{n}{6}$

$15 \times 6 = 4 \times n$

$\frac{90}{4} = \frac{4 \times n}{4}$

$22.5 = n$

Check:

$\frac{15}{4} \stackrel{?}{=} \frac{22.5}{6}$

$15 \times 6 \stackrel{?}{=} 4 \times 22.5$

$90 = 90$

25. $\frac{550}{n} = \frac{5}{3}$

$550 \times 3 = n \times 5$

$\frac{1650}{5} = \frac{n \times 5}{5}$

$330 = n$

Check:

$\frac{550}{330} \stackrel{?}{=} \frac{5}{3}$

$3 \times 550 \stackrel{?}{=} 330 \times 5$

$1650 = 1650$

27. $\frac{21}{n} = \frac{2}{3}$

$21 \times 3 = n \times 2$

$\frac{63}{2} = \frac{n \times 2}{2}$

$31.5 = n$

29. $\frac{9}{26} = \frac{n}{52}$

$9 \times 52 = 26 \times n$

$\frac{468}{26} = \frac{26 \times n}{26}$

$18 = n$

31. $\frac{15}{12} = \frac{10}{n}$

$15 \times n = 12 \times 10$

$\frac{15 \times n}{15} = \frac{120}{15}$

$n = 8$

33. $\frac{n}{36} = \frac{4.5}{1}$

$n \times 1 = 36 \times 4.5$

$n = 162$

35. $\frac{1.8}{n} = \frac{0.7}{12}$

$1.8 \times 12 = 0.7n$

$\frac{21.6}{0.7} = \frac{0.7n}{0.7}$

$30.9 \approx n$

37. $\frac{7}{8}=\frac{n}{4.2}$

$7\times 4.2=8\times n$

$\frac{29.4}{8}=\frac{8\times n}{8}$

$3.7\approx n$

39. $\frac{13.8}{15}=\frac{n}{6}$

$13.8\times 6=15\times n$

$82.8=15\times n$

$\frac{82.8}{15}=\frac{15\times n}{15}$

$5.5\approx n$

41. $\frac{3}{n}=\frac{6\frac{1}{4}}{100}$

$3\times 100=6\frac{1}{4}\times n$

$\frac{300}{\frac{25}{4}}=\frac{\frac{25}{4}\times n}{\frac{25}{4}}$

$48=n$

43. $\frac{n \text{ pounds}}{20 \text{ ounces}}=\frac{2 \text{ pounds}}{32 \text{ ounces}}$

$32n=20\times 2$

$\frac{32n}{32}=\frac{40}{32}$

$n=1.25$

45. $\frac{128 \text{ miles}}{4 \text{ hours}}=\frac{80 \text{ miles}}{n \text{ hours}}$

$128\times n=4\times 80$

$\frac{128\times n}{128}=\frac{320}{128}$

$n=2.5$

2.5 hours

47. $\frac{32 \text{ meters}}{5 \text{ yards}}=\frac{24 \text{ meters}}{n \text{ yards}}$

$32n=5\times 24$

$\frac{32n}{32}=\frac{120}{32}$

$n=3.75$

49. $\frac{3 \text{ inches}}{7.62 \text{ centimeters}}=\frac{n \text{ inches}}{10 \text{ centimeters}}$

$3\times 10=7.62n$

$\frac{30}{7.62}=\frac{7.62n}{7.62}$

$3.94\approx n$

51. $\frac{35 \text{ dimes}}{3.5 \text{ dollars}}=\frac{n \text{ dimes}}{8 \text{ dollars}}$

$35\times 8=3.5\times n$

$\frac{280}{3.5}=\frac{3.5\times n}{3.5}$

$80=n$

53. $\frac{3\frac{1}{4} \text{ feet}}{8 \text{ pounds}}=\frac{n \text{ feet}}{12 \text{ pounds}}$

$3\frac{1}{4}\times 12=8\times n$

$\frac{39}{8}=\frac{8\times n}{8}$

$4\frac{7}{8}=n$

55. $\frac{n}{5}=\frac{6.6}{3}$

$3n=5(6.6)$

$\frac{3n}{3}=\frac{33}{3}$

$n=11$ inches wide

57. $\frac{n}{2\frac{1}{3}}=\frac{4\frac{5}{6}}{3\frac{1}{9}}$

$\frac{28}{9}\times n=\frac{7}{3}\times\frac{29}{6}$

$\frac{\frac{28}{9}\times n}{\frac{28}{9}}=\frac{\frac{203}{18}}{\frac{28}{9}}$

$n=\frac{203}{18}\times\frac{9}{28}$

$n=\frac{28}{8}$ or $3\frac{5}{8}$

59. $\dfrac{8\frac{1}{6}}{n} = \dfrac{5\frac{1}{2}}{7\frac{1}{3}}$

$\dfrac{22}{3} \times \dfrac{49}{6} = \dfrac{11}{2} \times n$

$\dfrac{\frac{539}{9}}{\frac{11}{2}} = \dfrac{\frac{11}{2} \times n}{\frac{11}{2}}$

$\dfrac{539}{9} \times \dfrac{2}{11} = n$

$\dfrac{98}{9} = 10\dfrac{8}{9} = n$

Cumulative Review

61. $(3+1)^3 - 30 \div 6 - 144 \div 12$

$= 4^3 - 30 \div 6 - 144 \div 12$

$= 64 - 30 \div 6 - 144 \div 12$

$= 64 - 5 - 12$

$= 47$

63. 0.0034

65. $8 \times 7 = 56$ games

4.4 Exercises

1. He should continue with people on the top of the fraction. That would be 60 peole that he observed on Saturday night. He does not know the number of dogs, so this would be *n*. The proportion would be

$$\frac{12 \text{ people}}{5 \text{ dogs}} = \frac{60 \text{ people}}{n \text{ dogs}}$$

3. $\dfrac{140 \text{ sold}}{23 \text{ returned}} = \dfrac{980 \text{ sold}}{n \text{ returned}}$

$140 \times n = 23 \times 980$

$\dfrac{140 \times n}{n} = \dfrac{22,540}{140}$

$n = 161$

161 cars

5. $\dfrac{\frac{3}{4} \text{ cups}}{1 \text{ gallon}} = \dfrac{n \text{ cups}}{4 \text{ gallons}}$

$\dfrac{3}{4} \times 4 = 1 \times n$

$3 = n$

3 cups

7. $\dfrac{n \text{ kilometers}}{5 \text{ miles}} = \dfrac{1\frac{1}{2} \text{ kilometers}}{1 \text{ mile}}$

$n = 5 \times 1\dfrac{1}{2}$

$n = 7\dfrac{1}{2}$

$7\dfrac{1}{2}$ kilometers

9. $\dfrac{65 \text{ kronor}}{\$8 \text{ U.S.}} = \dfrac{n \text{ kronor}}{\$320 \text{ U.S.}}$

$65 \times 320 = 8 \times n$

$\dfrac{20,800}{8} = \dfrac{8 \times n}{8}$

$2600 = n$

2600 kronor

11. $\dfrac{6.5 \text{ feet}}{5 \text{ feet}} = \dfrac{n \text{ feet}}{152 \text{ feet}}$

$6.5 \times 152 = 5 \times n$

$\dfrac{988}{5} = \dfrac{5 \times n}{5}$

$197.6 = n$

197.6 feet

13. $\dfrac{3 \text{ inches}}{125 \text{ miles}} = \dfrac{5.2 \text{ inches}}{n \text{ miles}}$

$3 \times n = 125 \times 5.2$

$\dfrac{3 \times n}{3} = \dfrac{650}{3}$

$n = 216.\overline{6}$

217 miles

15. $\frac{2 \text{ cups}}{7 \text{ people}} = \frac{n \text{ cups}}{62 \text{ people}}$

$$2 \times 62 = 7 \times n$$
$$\frac{124}{7} = \frac{7 \times n}{7}$$
$$17.7 \approx n$$

17.7 cups

17. $\frac{17 \text{ made}}{25 \text{ free throws}} = \frac{n \text{ made}}{150 \text{ free throws}}$

$$17 \times 150 = 25 \times n$$
$$\frac{2250}{25} = \frac{25 \times n}{25}$$
$$102 = n$$

102 free throws

19. $\frac{n \text{ gallons}}{600 \text{ miles}} = \frac{6 \text{ gallons}}{192 \text{ miles}}$

$$192 \times n = 600 \times 6$$
$$\frac{192 \times n}{192} = \frac{3600}{192}$$
$$n = 18.75$$

18.75 gallons

21. $\frac{24 \text{ tagged}}{n \text{ total}} = \frac{12 \text{ tagged}}{20 \text{ total}}$

$$24 \times 20 = 12 \times n$$
$$\frac{480}{12} = \frac{12 \times n}{12}$$
$$40 = n$$

40 hawks

23. $\frac{425 \text{ pounds}}{3 \text{ acres}} = \frac{n \text{ pounds}}{14 \text{ acres}}$

$$425 \times 14 = 3 \times n$$
$$\frac{5950}{3} = \frac{3 \times n}{3}$$
$$1983.\overline{3} = n$$

$1983\frac{1}{3}$ pounds

$$1983\frac{1}{3} \times 1.8 = 3570$$

$3570

25. $\frac{5 \text{ defective}}{100 \text{ made}} = \frac{n \text{ defective}}{5400 \text{ made}}$

$$5 \times 5400 = 100 \times n$$
$$\frac{27{,}000}{100} = \frac{100 \times n}{n}$$
$$270 = n$$

270 defective chips

27. Water: $\frac{n \text{ cups}}{3 \text{ servings}} = \frac{2 \text{ cups}}{6 \text{ servings}}$

$$n \times 6 = 3 \times 2$$
$$\frac{n \times 6}{6} = \frac{6}{6}$$
$$n = 1$$

1 cup of water

Milk: $\frac{n \text{ cups}}{3 \text{ servings}} = \frac{\frac{3}{4} \text{ cups}}{6 \text{ servings}}$

$$n \times 6 = 3 \times \frac{3}{4}$$
$$n \times 6 = \frac{9}{4}$$
$$\frac{n \times 6}{6} = \frac{\frac{9}{4}}{6}$$
$$n = \frac{9}{4} \div 6$$
$$= \frac{9}{4} \times \frac{1}{6} = \frac{3}{8}$$

$\frac{3}{8}$ cup of milk

1 cup of water and 3/8 cup of milk

29. Water: $\frac{n \text{ cups}}{8 \text{ servings}} = \frac{2 \text{ cups}}{6 \text{ servings}}$

$$n \times 6 = 8 \times 2$$
$$\frac{n \times 6}{6} = \frac{16}{6}$$
$$n = \frac{16}{6} = \frac{8}{3}$$

High altitude: $\frac{8}{3} \times \frac{3}{4} = 2$

2 cups of water

Milk: $\frac{n \text{ cups}}{8 \text{ servings}} = \frac{\frac{3}{4} \text{ cups}}{6 \text{ servings}}$

$$n \times 6 = 8 \times \frac{3}{4}$$
$$n \times 6 = 6$$
$$\frac{n \times 6}{6} = \frac{6}{6}$$
$$n = 1$$

1 cup of milk

2 cups of water and 1 cup of milk

31. Rodriguez:

$\frac{\$22,000,000}{57 \text{ home runs}} = \$385,965 / \text{home run}$

Sosa:

$\frac{\$15,000,000}{49 \text{ home runs}} = \$206,122 / \text{home run}$

Alex Rodriguez, approximately $385,965 for each home run; Sammy Sosa, approximately $306,122 for each home run.

33. Johnson:

$\frac{\$13,350,000}{35 \text{ games}} = \$381,429 / \text{game}$

Mussina:

$\frac{\$11,000,000}{33 \text{ games}} = \$333,333 / \text{game}$

Cumulative Review

35. 56,179 rounds to 56,200

37. 56.148 rounds to 56.1

39. a. $1\frac{3}{16} \times \frac{4}{5} = \frac{19}{16} \times \frac{4}{5} = \frac{76}{80}$

$$= \frac{19}{20} \text{ of a square foot}$$

b. $\frac{19}{20} \times 1500 = 19 \times 75$

$$= 1452 \text{ square feet}$$

Putting Your Skills To Work

1. a. $180 + 70 = 250$ $\quad$ $250 \times 5 = 1250$

1250 calories

b. $\frac{n \text{ hours}}{1250 \text{ calories}} = \frac{1 \text{ hour}}{250 \text{ calories}}$

$$250 \times n = 1250$$
$$\frac{250 \times n}{250} = \frac{1250}{250}$$
$$n = 5$$

5 hours

2. a. $590 + 450 = 1040$ $\quad$ $1040 \times 5 = 5200$

5200 calories

b. $\frac{n \text{ hours}}{5200 \text{ calories}} = \frac{1 \text{ hour}}{200 \text{ calories}}$

$$20 \times n = 1 \times 5200$$
$$\frac{200 \times n}{200} = \frac{5200}{200}$$
$$n = 26$$

26 hours

3. Nora; $2\times 4 = 8$ times

$$\begin{array}{r} 560 \\ \times\ 8 \\ \hline 4480 \end{array}$$ calories/month

Gina; $5\times 4 = 20$ times

$$\begin{array}{r} 230 \\ \times\ 20 \\ \hline 4600 \end{array}$$

Gina consumes more calories

$$\begin{array}{r} 4600 \\ -\ 4480 \\ \hline 120 \end{array}$$ calories more

4. Barry;

$$\begin{array}{r} 250 \\ \times\ 4 \\ \hline 1000 \end{array} \quad \begin{array}{r} 350 \\ \times\ 3 \\ \hline 1050 \end{array} \quad \begin{array}{r} 1050 \\ +\ 1000 \\ \hline 2050 \end{array}$$ calories

Miguel;

$$\begin{array}{r} 500 \\ \times\ 2 \\ \hline 1000 \end{array} \quad \begin{array}{r} 700 \\ \times\ 2 \\ \hline 1400 \end{array} \quad \begin{array}{r} 1000 \\ +\ 1400 \\ \hline 2400 \end{array}$$ calories

Miguel burns more calories

$$\begin{array}{r} 2400 \\ -\ 2050 \\ \hline 350 \end{array}$$ more calories

5. $5\times 4 = 20$ $$\begin{array}{r} 260 \\ \times\ 20 \\ \hline 5200 \end{array}$$

Answers may vary.

Sample answer: 2 hours tae kwon do, 5 hours cleaning house, 6 hours jogging

6. a. $\frac{700}{500} = \frac{7}{5}$

b. $\frac{350}{250} = \frac{7}{5}$

Chapter 4 Review Problems

1. $88:40 = \frac{88}{40} = \frac{88\div 8}{40\div 8} = \frac{11}{5}$

2. $65:39 = \frac{65}{39} = \frac{65\div 13}{39\div 13} = \frac{5}{3}$

3. $28:35 = \frac{28}{35} = \frac{28\div 7}{35\div 7} = \frac{4}{5}$

4. $250:475 = \frac{250}{475} = \frac{250\div 25}{475\div 25} = \frac{10}{19}$

5. $2\frac{1}{3}$ to $4\frac{1}{4} = \frac{2\frac{1}{3}}{4\frac{1}{4}} = 2\frac{1}{3}\div 4\frac{1}{4}$

$= \frac{7}{3}\div\frac{17}{4} = \frac{7}{3}\times\frac{4}{17}$

$= \frac{28}{51}$

6. $\frac{27}{81} = \frac{27\div 27}{81\div 27} = \frac{1}{3}$

7. $\frac{280}{651} = \frac{280\div 7}{651\div 7} = \frac{40}{93}$

8. $\frac{156}{441} = \frac{156\div 3}{441\div 3} = \frac{52}{147}$

9. $\frac{26}{65} = \frac{26\div 13}{65\div 13} = \frac{2}{5}$

10. $\frac{\$60}{\$480} = \frac{60\div 60}{480\div 60} = \frac{1}{8}$

11. $$\begin{array}{r} 60 \\ +\ 45 \\ \hline 105 \end{array}$$ $\frac{\$105}{\$480} = \frac{105\div 15}{480\div 15} = \frac{7}{32}$

12. $\frac{\$75}{6 \text{ people}} = \frac{\$25}{2 \text{ people}}$

13. $\frac{44 \text{ revolutions}}{121 \text{ minutes}} = \frac{4 \text{ revolutions}}{11 \text{ minutes}}$

14. $\frac{188 \text{ vibrations}}{16 \text{ seconds}} = \frac{47 \text{ vibrations}}{4 \text{ seconds}}$

15. $\frac{20 \text{ cups}}{32 \text{ pies}} = \frac{5 \text{ cups}}{8 \text{ pies}}$

16. $\frac{\$2125}{125 \text{ shares}} = \frac{\$17}{1 \text{ share}} = \$17/\text{share}$

17. $\frac{\$1344}{12 \text{ credit-hours}} = \$112/\text{credit-hour}$

18. $\frac{\$742.50}{55 \text{ sq yd}} = \frac{\$13.50}{1 \text{ yd}} = \$13.50/\text{square yard}$

19. $\frac{\$600}{48 \text{ DVD}} = \$12.50/\text{DVD}$

20. a. $\frac{\$2.96}{4 \text{ oz}} = \frac{\$2.96 \div 4}{4 \text{ oz} \div 4} = \$0.74/\text{oz}$

b. $\frac{\$5.22}{9 \text{ oz}} = \frac{\$5.22 \div 9}{9 \text{ oz} \div 9} = \$0.58/\text{oz}$

c. $\$0.74 - 0.58 = \$0.16/\text{oz}$

21. a. $\frac{\$2.75}{12.5 \text{ oz}} = \frac{\$2.75 \div 12.5}{12.5 \text{ oz} \div 12.5} = \$0.22/\text{oz}$

b. $\frac{\$1.75}{7.0 \text{ oz}} = \frac{\$1.75 \div 7.0}{7.0 \text{ oz} \div 7.0} = \$0.25/\text{oz}$

c. $0.25 - 0.22 = 0.03 = \$0.03/\text{oz}$

22. $\frac{12}{48} = \frac{7}{28}$

23. $\frac{1\frac{1}{2}}{5} = \frac{4}{13\frac{1}{3}}$

24. $\frac{7.5}{45} = \frac{22.5}{135}$

25. $\frac{138 \text{ passengers}}{3 \text{ buses}} = \frac{230 \text{ passengers}}{5 \text{ buses}}$

26. $\frac{\$4.50}{15 \text{ pounds}} = \frac{\$8.10}{27 \text{ pounds}}$

27. $\frac{16}{48} \stackrel{?}{=} \frac{2}{12}$

$16 \times 12 \stackrel{?}{=} 48 \times 2$

$192 \neq 96$

It is not a proportion.

28. $\frac{20}{25} \stackrel{?}{=} \frac{8}{10}$

$25 \times 10 \stackrel{?}{=} 25 \times 8$

$200 = 200$

It is a proportion.

29. $\frac{24}{20} \stackrel{?}{=} \frac{18}{15}$

$24 \times 15 \stackrel{?}{=} 20 \times 18$

$360 = 360$

It is a proportion.

30. $\frac{84}{48} \stackrel{?}{=} \frac{14}{8}$

$84 \times 8 \stackrel{?}{=} 48 \times 14$

$672 = 672$

It is a proportion.

31. $\frac{37}{33} \stackrel{?}{=} \frac{22}{19}$

$37 \times 19 \stackrel{?}{=} 33 \times 22$

$703 \neq 726$

It is not a proportion.

32. $\frac{15}{18} \stackrel{?}{=} \frac{18}{22}$

$15 \times 22 \stackrel{?}{=} 18 \times 18$

$330 \neq 324$

It is not a proportion.

33. $\frac{84 \text{ miles}}{7 \text{ gallons}} \stackrel{?}{=} \frac{108 \text{ miles}}{9 \text{ gallons}}$

$84 \times 9 \stackrel{?}{=} 7 \times 108$

$756 = 756$

It is a proportion.

34. $\frac{156 \text{ rev}}{6 \text{ min}} \stackrel{?}{=} \frac{181 \text{ rev}}{7 \text{ min}}$

$156 \times 7 \stackrel{?}{=} 6 \times 181$

$1092 \neq 1086$

It is not a proportion.

35. $7 \times n = 161$

$\frac{7 \times n}{7} = \frac{161}{7}$

$n = 23$

36. $8 \times n = 42$

$\frac{8 \times n}{8} = \frac{42}{8}$

$n = 5\frac{1}{4} = 5.25$

37. $442 = 20 \times n$

$\frac{442}{20} = \frac{20 \times n}{20}$

$22.1 = n$

38. $663 = 39 \times n$

$\frac{663}{39} = \frac{39 \times n}{39}$

$17 = n$

39. $\frac{3}{11} = \frac{9}{n}$

$3 \times n = 11 \times 9$

$\frac{3 \times n}{3} = \frac{99}{3}$

$n = 33$

40. $\frac{2}{7} = \frac{12}{n}$

$2 \times n = 7 \times 12$

$\frac{2 \times n}{2} = \frac{84}{2}$

$n = 42$

41. $\frac{n}{28} = \frac{6}{24}$

$24 \times n = 28 \times 6$

$\frac{24 \times n}{24} = \frac{168}{24}$

$n = 7$

42. $\frac{n}{32} = \frac{15}{20}$

$n \times 20 = 32 \times 15$

$\frac{n \times 20}{20} = \frac{480}{20}$

$n = 24$

43. $\frac{2\frac{1}{4}}{9} = \frac{4\frac{3}{4}}{n}$

$2\frac{1}{4} \times n = 9 \times 4\frac{3}{4}$

$\frac{9}{4} \times n = 9 \times \frac{19}{4}$

$\frac{\frac{9}{4} \times n}{\frac{9}{4}} = \frac{\frac{171}{4}}{\frac{9}{4}}$

$n = \frac{171}{4} \div \frac{9}{4}$

$= \frac{171}{4} \times \frac{4}{9}$

$= \frac{171}{9}$

$= 19$

44. $\frac{3\frac{1}{3}}{2\frac{2}{3}} = \frac{7}{n}$

$3\frac{1}{3} \times n = 2\frac{2}{3} \times 7$

$\frac{10}{3} \times n = \frac{8}{3} \times 7$

$\frac{\frac{10}{3} \times n}{\frac{10}{3}} = \frac{\frac{56}{3}}{\frac{10}{3}}$

$n = \frac{56}{3} \div \frac{10}{3}$

$= \frac{56}{3} \times \frac{3}{10}$

$= \frac{56}{10} = \frac{28}{5}$

$= 5\frac{3}{5} = 5.6$

45. $\frac{42}{50} = \frac{n}{6}$

$6 \times 42 = 50 \times n$

$\frac{252}{50} = \frac{50 \times n}{50}$

$5.0 \approx n$

46. $\frac{38}{45} = \frac{n}{8}$

$8 \times 38 = 45 \times n$

$\frac{304}{45} = \frac{45 \times n}{45}$

$6.8 \approx n$

47. $\frac{2.25}{9} = \frac{4.75}{n}$

$2.25 \times n = 9 \times 4.75$

$\frac{2.25 \times n}{2.25} = \frac{42.75}{2.25}$

$n = 19$

48. $\frac{3.5}{5} = \frac{10.5}{n}$

$3.5 \times n = 5(10.5)$

$\frac{3.5 \times n}{3.5} = \frac{52.5}{3.5}$

$n = 15$

49. $\frac{25}{7} = \frac{60}{n}$

$25 \times n = 7 \times 60$

$\frac{25 \times n}{25} = \frac{420}{25}$

$n = 16.8$

50. $\frac{60}{9} = \frac{31}{n}$

$60 \times n = 9 \times 31$

$60 \times n = 279$

$\frac{60 \times n}{60} = \frac{279}{60}$

$n \approx 4.7$

51. $\frac{35 \text{ miles}}{28 \text{ gallons}} = \frac{15 \text{ miles}}{n \text{ gallons}}$

$35 \times n = 28 \times 15$

$\frac{35 \times n}{35} = \frac{420}{35}$

$n = 12$

52. $\frac{8 \text{ defective}}{100 \text{ perfect}} = \frac{44 \text{ defective}}{n \text{ perfect}}$

$8 \times n = 100 \times 44$

$\frac{8 \times n}{8} = \frac{4400}{8}$

$n = 550$

550 perfect parts

53. $\frac{3 \text{ gallons}}{2 \text{ rooms}} = \frac{n \text{ gallons}}{10 \text{ rooms}}$

$3 \times 10 = 2 \times n$

$\frac{30}{2} = \frac{2 \times n}{2}$

$15 = n$

15 gallons

54. $\frac{49 \text{ coffee}}{100 \text{ adults}} = \frac{n \text{ coffee}}{3450 \text{ adults}}$

$49 \times 3450 = 100 \times n$

$\frac{169{,}050}{100} = \frac{100 \times n}{100}$

$1691 \approx n$

1691 employees drink coffee

55. $\frac{24 \text{ francs}}{5 \text{ dollars}} = \frac{n \text{ francs}}{420 \text{ dollars}}$

$24 \times 420 = 5 \times n$

$\frac{10{,}080}{5} = \frac{5 \times n}{5}$

$2016 = n$

2016 francs

56. $\frac{n \text{ francs}}{\$80} = \frac{5 \text{ francs}}{\$3.6}$

$3.6 \times n = 80 \times 5$

$\frac{3.6 \times n}{3.6} = \frac{400}{3.6}$

$n \approx 111.11$

111.11 Swiss francs

57. $\frac{225 \text{ miles}}{3 \text{ inches}} = \frac{n \text{ miles}}{8 \text{ inches}}$

$8 \times 225 = 3 \times n$

$\frac{1800}{3} = \frac{3 \times n}{3}$

$600 = n$

600 miles

58. $\frac{n \text{ free throws}}{10 \text{ games}} = \frac{14 \text{ free throws}}{4 \text{ games}}$

$4 \times n = 10 \times 14$

$\frac{4 \times n}{4} = \frac{140}{4}$

$n = 35$

35 free throws

59. $\frac{6 \text{ feet}}{16 \text{ feet}} = \frac{n \text{ feet}}{320 \text{ feet}}$

$6 \times 320 = 16 \times n$

$\frac{1920}{16} = \frac{16 \times n}{16}$

$120 = n$

120 feet

60. a. $\frac{680 \text{ miles}}{26 \text{ gallons}} = \frac{200 \text{ miles}}{n \text{ gallons}}$

$680 \times n = 26 \times 200$

$\frac{680 \times n}{680} = \frac{5200}{680}$

$n \approx 7.65$ gallons

b.

$$\begin{array}{r} 7.65 \\ \times\ 1.85 \\ \hline 3825 \\ 6120 \\ 765 \\ \hline 14.1525 \end{array}$$

\$14.15

61. $\frac{3.5 \text{ cm}}{2.5 \text{ cm}} = \frac{8 \text{ cm}}{n \text{ cm}}$

$3.5 \times n = 2.5 \times 8$

$\frac{3.5 \times n}{3.5} = \frac{20}{3.5}$

$n \approx 5.71$

5.71 centimeters tall

62. $\frac{3 \text{ grams}}{50 \text{ pounds}} = \frac{n \text{ grams}}{125 \text{ pounds}}$

$3 \times 125 = 50 \times n$

$\frac{375}{50} = \frac{50 \times n}{50}$

$7.5 = n$

7.5 grams

63. $\frac{33 \text{ bricks}}{2 \text{ feet}} = \frac{n \text{ bricks}}{11 \text{ feet}}$

$33 \times 11 = 2 \times n$

$\frac{363}{2} = \frac{2 \times n}{2}$

$181.5 = n$

She needs 181.5 bricks which equals 182 bricks.

64. $\frac{n \text{ pages}}{6 \text{ hours}} = \frac{20 \text{ pages}}{2.5 \text{ hours}}$

$2.5 \times n = 16 \times 20$

$\frac{2.5 \times n}{2.5} = \frac{120}{2.5}$

$n = 48$

48 pages

65. $\frac{3 \text{ gallons}}{500 \text{ sq ft}} = \frac{n \text{ gallons}}{1400 \text{ sq ft}}$

$3 \times 1400 = 500 \times n$

$\frac{4200}{500} = \frac{500 \times n}{500}$

$8.4 = n$

He needs 8.4 gallons which equals 9 gallons.

66. $\frac{n \text{ liters}}{1250 \text{ runners}} = \frac{2 \text{ liters}}{3 \text{ runners}}$

$3 \times n = 1250 \times 2$

$\frac{3 \times n}{3} = \frac{2500}{3}$

$n \approx 833.33$

834 liters

67. $\frac{14 \text{ cm}}{145 \text{ feet}} = \frac{11 \text{ cm}}{n \text{ feet}}$

$14 \times n = 145 \times 11$

$\frac{14 \times n}{14} = \frac{1595}{14}$

$n \approx 113.93$

Width is 113.93 feet

68. $\frac{4 \text{ days}}{3 \text{ minutes}} = \frac{365 \text{ days}}{n \text{ minutes}}$

$40 \times n = 3 \times 365$

$\frac{40 \times n}{40} = \frac{1095}{40}$

$n \approx 27.38$

27.38 minutes

69. $\frac{68 \text{ goals}}{27 \text{ games}} = \frac{n \text{ goals}}{34 \text{ games}}$

$68 \times 34 = 27 \times n$

$\frac{2312}{27} = \frac{27 \times n}{27}$

$86 \approx n$

86 goals

70. $\frac{345 \text{ calories}}{10 \text{ ounces}} = \frac{n \text{ calories}}{16 \text{ ounces}}$

$345 \times 16 = 10 \times n$

$\frac{5520}{10} = \frac{10 \times n}{10}$

$552 = n$

71. $\frac{n \text{ people}}{45{,}600 \text{ residents}} = \frac{3 \text{ people}}{10 \text{ residents}}$

$10 \times n = 45{,}600 \times 3$

$\frac{10 \times n}{10} = \frac{136{,}800}{10}$

$n \approx 13{,}680$

13,680 people

72. $\frac{n \text{ trips}}{240 \text{ trips}} = \frac{13 \text{ trips}}{16 \text{ trips}}$

$16 \times n = 240 \times 13$

$\frac{16 \times n}{16} = \frac{3120}{16}$

$n = 195$

195 trips

How Am I Doing? Chapter 4 Test

1. $\frac{18}{52}=\frac{18\div 2}{52\div 2}=\frac{9}{26}$

2. $\frac{70}{185}=\frac{70\div 5}{185\div 5}=\frac{14}{37}$

3. $\frac{784 \text{ miles}}{24 \text{ gal}}=\frac{784 \text{ miles}\div 8}{24 \text{ gal}\div 8}=\frac{98 \text{ miles}}{3 \text{ gal}}$

4. $\frac{2100 \text{ sq ft}}{45 \text{ lb}}=\frac{2100 \text{ sq ft}\div 15}{45 \text{ lb}\div 15}=\frac{140 \text{ sq ft}}{3 \text{ lb}}$

5. $\frac{19 \text{ tons}}{5 \text{ days}}=\frac{19 \text{ tons}\div 5}{5 \text{ days}\div 5}=3.8 \text{ tons/day}$

6. $\frac{\$57.96}{7 \text{ hr}}=\frac{\$57.96\div 7}{7 \text{ hr}\div 7}=\$8.28/\text{hr}$

7. $\frac{5400 \text{ ft}}{22\text{poles}}=\frac{5400 \text{ ft}\div 22}{22 \text{ poles}\div 22}=245.45 \text{ ft/pole}$

8. $\frac{\$9373}{110 \text{ shares}}=\frac{\$9373\div 110}{110 \text{ shares}\div 110}$
$=\$85.21/\text{share}$

9. $\frac{17}{29}=\frac{51}{87}$

10. $\frac{2\frac{1}{2}}{10}=\frac{6}{24}$

11. $\frac{490 \text{ miles}}{21 \text{ gallons}}=\frac{280 \text{ miles}}{12 \text{ gallons}}$

12. $\frac{3 \text{ hours}}{180 \text{ miles}}=\frac{5 \text{ hours}}{300 \text{ miles}}$

13. $\frac{50}{24}\overset{?}{=}\frac{35}{16}$

$50\times 16 \overset{?}{=} 24\times 34$

$1800 \neq 816$

It is not a proportion.

14. $\frac{3\frac{1}{2}}{14}=\frac{5}{20}$

$3\frac{1}{2}\times 20 \overset{?}{=} 14\times 5$

$\frac{7}{2}\times\frac{20}{1}=70$

$70=70$

It is a proportion.

15. $\frac{32 \text{ smokers}}{46 \text{ nonsmokers}}\overset{?}{=}\frac{160 \text{ smokers}}{230 \text{ nonsmokers}}$

$32\times 230 \overset{?}{=} 46\times 160$

$7360 \neq 7360$

It is not a proportion.

16. $\frac{\$0.74}{16 \text{ oz}}\overset{?}{=}\frac{\$1.84}{40 \text{ oz}}$

$0.74\times 40 \overset{?}{=} 16\times 1.84$

$29.6 \neq 29.44$

It is not a proportion.

17. $\frac{n}{20}=\frac{4}{5}$

$n\times 5=20\times 4$

$\frac{n\times 5}{5}=\frac{80}{5}$

$n=16$

18. $\frac{8}{3}=\frac{60}{n}$

$8\times n=3\times 60$

$\frac{8\times n}{8}=\frac{180}{8}$

$n=22.5$

19. $\frac{2\frac{2}{3}}{8} = \frac{6\frac{1}{3}}{n}$

$2\frac{2}{3} \times n = 8 \times 6\frac{1}{3}$

$\frac{8}{3} \times n = 8 \times \frac{19}{3}$

$\frac{\frac{8}{3} \times n}{\frac{8}{3}} = \frac{\frac{152}{3}}{\frac{8}{3}}$

$n = \frac{152}{3} \div \frac{152}{3} \times \frac{3}{8}$

$= 19$

20. $\frac{4.2}{11} = \frac{n}{77}$

$4.2 \times 77 = 11 \times n$

$\frac{323.4}{11} = \frac{11 \times n}{11}$

$29.4 = n$

21. $\frac{45 \text{ women}}{15 \text{ men}} = \frac{n \text{ women}}{40 \text{ men}}$

$45 \times 40 = 15 \times n$

$\frac{1800}{15} = \frac{15 \times n}{15}$

$120 = n$

120 women

22. $\frac{5 \text{ kg}}{11 \text{ pounds}} = \frac{32 \text{ kg}}{n \text{ pounds}}$

$5 \times n = 11 \times 36$

$\frac{5 \times n}{5} = \frac{396}{5}$

$n = 70.5$

70.5 pounds

23. $\frac{n \text{ inches of snow}}{14 \text{ inches of rain}} = \frac{12 \text{ inches of snow}}{1.4 \text{ inches of rain}}$

$n \times 1.4 = 14 \times 12$

$\frac{n \times 1.4}{1.4} = \frac{168}{1.4}$

$n = 120$

120 inches of snow

24. $\frac{5 \text{ pounds}}{\$n} = \frac{\frac{1}{2} \text{ pounds}}{\$5.20}$

$5 \times 5.20 = \frac{1}{2} \times n$

$\frac{2600}{\frac{1}{2}} = \frac{\frac{1}{2} \times n}{\frac{1}{2}}$

$52 = n$

\$52

25. $\frac{3 \text{ eggs}}{11 \text{ people}} = \frac{n \text{ eggs}}{22 \text{ people}}$

$3 \times 22 = 11 \times n$

$\frac{66}{11} = \frac{11 \times n}{11}$

$6 = n$

6 eggs

26. $\frac{42 \text{ ft}}{170 \text{ lb}} = \frac{20 \text{ ft}}{n \text{ lb}}$

$42 \times n = 170 \times 20$

$\frac{42 \times n}{42} = \frac{3400}{42}$

$n \approx 80.95$

80.95 pounds

27. $\frac{9 \text{ inches}}{57 \text{ miles}} = \frac{3 \text{ inches}}{n \text{ miles}}$

$9 \times n = 57 \times 3$

$\frac{9 \times n}{9} = \frac{171}{9}$

$n = 19$

19 miles

28. $\frac{\$240}{4000 \text{ sq ft}} = \frac{\$n}{6000 \text{ sq ft}}$

$240 \times 6000 = 4000 \times n$

$\frac{1{,}440{,}000}{4000} = \frac{4000 \times n}{4000}$

$360 = n$

\$360

29. $\frac{n \text{ miles}}{220 \text{ km}} = \frac{1 \text{ mile}}{1.61 \text{ km}}$

$1.61 \times n = 220 \times 1$

$\frac{1.61 \times n}{1.61} = \frac{220}{1.61}$

$n \approx 136.646$

136.7 miles

30. $\frac{570 \text{ km}}{9 \text{ hr}} = \frac{n \text{ km}}{11 \text{ hr}}$

$570 \times 11 = 9 \times n$

$6270 = 9 \times n$

$\frac{6270}{9} = \frac{9 \times n}{9}$

$696.67 \approx n$

696.67 km

31. $\frac{n \text{ free throws made}}{120 \text{ free throws}} = \frac{11 \text{ free throws made}}{15 \text{ free throws}}$

$15 \times n = 120 \times 11$

$\frac{15 \times n}{15} = \frac{1320}{15}$

$n = 88$

88 more free throws made

32. $\frac{7 \text{ hits}}{34 \text{ bats}} = \frac{n \text{ hits}}{155 \text{ bats}}$

$7 \times 155 = 34 \times n$

$\frac{1085}{34} = \frac{34 \times n}{34}$

$32 \approx n$

32 hits

Cumulative Test for Chapters 1-4

1. $26,597,089$ = Twenty-six million, five hundred ninety-seven thousand, eighty nine

2.

$$\begin{array}{r} 68 \\ 23\overline{)1564} \\ \underline{138} \\ 184 \\ \underline{184} \\ 0 \end{array}$$

3. $\frac{1}{4} + \frac{1}{8} \times \frac{3}{4} = \frac{1}{4} = \frac{3}{24}$

$= \frac{1}{4} \times \frac{8}{8} + \frac{3}{24}$

$= \frac{8}{32} + \frac{3}{32} = \frac{11}{32}$

4. $2\frac{1}{2} - 1\frac{3}{7} = 2\frac{7}{35} - 1\frac{15}{35}$

$= 1\frac{42}{35} - 1\frac{15}{35}$

$= \frac{27}{35}$

5. $\frac{20}{1} \times 3\frac{1}{4} = \frac{20}{1} \times \frac{13}{4} = 65$

6.

$$\begin{array}{r} 12,100 \\ -\ 3.8416 \\ \hline 8.2584 \end{array}$$

7.

$$\begin{array}{r} 0.8163 \\ \times\ \ 0.22 \\ \hline 16326 \\ \underline{16326\ } \\ 0.179586 \end{array}$$

8. $\frac{5}{12} \div \frac{15}{6} = \frac{5}{12} \times \frac{6}{15} = \frac{1}{6}$

9. $16.1455 \times 1000 = 16{,}145.5$

10. 56.<u>8</u>918 rounds to 56.9

11.
$$\begin{array}{r} 258.920 \\ +\quad 67.358 \\ \hline 326.278 \end{array}$$

12.
$$\begin{array}{r} 3.68 \\ 0.15\overline{)0.5520} \\ \underline{45} \\ 102 \\ \underline{90} \\ 120 \\ \underline{120} \\ 0 \end{array}$$

13.
$$\begin{array}{r} .15625 \\ 32\overline{)5.0000} \\ \underline{32} \\ 180 \\ \underline{160} \\ 200 \\ \underline{192} \\ 80 \\ \underline{64} \\ 160 \end{array}$$

$\frac{5}{32} = 0.15625$

14. $\frac{12}{17} \stackrel{?}{=} \frac{30}{42.5}$

$12 \times 42.5 \stackrel{?}{=} 17 \times 30$

$510 = 510$

It is a proportion.

15. $\frac{4\frac{1}{3}}{13} \stackrel{?}{=} \frac{2\frac{2}{3}}{8}$

$4\frac{1}{3} \times 8 \stackrel{?}{=} 13 \times 2\frac{2}{3}$

$\frac{13}{3} \times 8 \stackrel{?}{=} 13 \times \frac{8}{3}$

$\frac{104}{3} = \frac{104}{3}$

It is a proportion.

16. $\frac{9}{2.1} = \frac{n}{0.7}$

$9 \times 0.7 = 2.1 \times n$

$\frac{6.3}{2.1} = \frac{2.1 \times n}{2.1}$

$3 = n$

17. $\frac{50}{20} = \frac{5}{n}$

$50 \times n = 20 \times 5$

$\frac{50 \times n}{50} = \frac{100}{50}$

$n = 2$

18. $\frac{n}{56} = \frac{16}{14}$

$14 \times n = 56 \times 16$

$\frac{14 \times n}{14} = \frac{896}{14}$

$n = 64$

19. $\frac{7}{n} = \frac{28}{36}$

$7 \times 36 = n \times 28$

$\frac{252}{28} = \frac{n \times 28}{28}$

$n = 9$

20. $\frac{n}{11} = \frac{5}{16}$

$n \times 16 = 11 \times 5$

$\frac{n \times 16}{16} = \frac{55}{16}$

$n \approx 3.4$

21. $\frac{3\frac{1}{3}}{7} = \frac{10}{n}$

$3\frac{1}{3} \times n = 7 \times 10$

$\frac{\frac{10}{3} \times n}{\frac{10}{3}} = \frac{70}{\frac{10}{3}}$

$n = 70 \times \frac{3}{10}$

$n = 21$

22. $\frac{300 \text{ miles}}{4 \text{ inches}} = \frac{625 \text{ miles}}{n \text{ inches}}$

$300 \times n = 4 \times 625$

$\frac{300 \times n}{300} = \frac{2500}{300}$

$n \approx 8.33$

8.33 inches

23. $\frac{\$n}{35 \text{ hours}} = \frac{\$34}{3 \text{ hours}}$

$3 \times n = 35 \times 34$

$\frac{3 \times n}{3} = \frac{1190}{3}$

$n = 396.67$

$396.67

24. $\frac{14 \text{ people}}{3.5 \text{ lb}} = \frac{20 \text{ people}}{n \text{ lb}}$

$14 \times n = 3.5 \times 20$

$\frac{14 \times n}{14} = \frac{70}{14}$

$n = 5$

5 pounds

25. $\frac{39 \text{ gallons sap}}{2 \text{ gallons syrup}} = \frac{n \text{ gallons sap}}{11 \text{ gallons syrup}}$

$39 \times 11 = 2 \times n$

$\frac{429}{2} = \frac{2 \times n}{2}$

$214.5 = n$

214.5 gallons of maple sap

Chapter 5

5.1 Exercises

1. hundred

3. Move the decimal point two places to the left. Drop the % symbol.

5. $\frac{59}{100} = 59\%$

7. $\frac{4}{100} = 4\%$

9. $\frac{80}{100} = 80\%$

11. $\frac{245}{100} = 245\%$

13. $\frac{5.3}{100} = 5.3\%$

15. $\frac{0.07}{100} = 0.07\%$

17. $\frac{13}{100} = 13\%$

19. $\frac{8}{100} = 8\%$

21. $51\% = 0.51$

23. $7\% = 0.07$

25. $20\% = 0.20 = 0.2$

27. $43.6\% = 0.436$

29. $0.03\% = 0.0003$

31. $0.72\% = 0.0072$

33. $1.25\% = 0.0125$

35. $366\% = 3.66$

37. $0.74\% = 74\%$

39. $0.50 = 50\%$

41. $0.08 = 8\%$

43. $0.563 = 56.3\%$

45. $0.002 = 0.2\%$

47. $0.0057 = 0.57\%$

49. $1.35 = 135\%$

51. $2.72 = 272\%$

53. $0.27 = 27\%$

55. $0.3 = 30\%$

57. $0.94 = 94\%$

59. $2.31 = 231\%$

61. $\frac{10}{100} = 10\%$

63. $0.009 = 0.9\%$

65. $62\% = \frac{62}{100} = 0.62$

67. $138\% = \frac{138}{100} = 1.38$

69. $\frac{0.3}{100} = 0.003$

71. $\frac{80}{100} = 0.8$

73. $\frac{59}{100} = 59\%$

75. $115\% = 1.15$

77. $0.6\% = 0.006$

79. Starbucks: $77\% = 0.77$
Krispy Kreme: $90\% = 0.9$

81. 36% = 36 percent = 36 "per one hundred" =
$36 \times \frac{1}{100} = \frac{36}{100} = 0.36$. The rule is using the fact that 36% means 36 per one hundred.

83. a. $1562\% = 15.62$
b. $1562\% = \frac{1562}{100}$
c. $1562\% = \frac{1562}{100} = \frac{781}{50}$

Cumulative Review

85. $0.56 = \frac{56}{100} = \frac{14}{25}$

87.
$$\begin{array}{r} 0.6875 \\ 16\overline{)11.000} \\ \underline{96} \\ 140 \\ \underline{128} \\ 120 \\ \underline{112} \\ 80 \end{array}$$
$\frac{11}{16} = 0.6875$

89. $3 \times 246 + 7 \times 380 + 5 \times 168 + 9 \times 122$
$= 738 + 2660 + 840 + 1098$
$= 5336$ vases

5.2 Exercises

1. Write the number in front of the percent symbol as the numerator of the fraction. Write the number 100 as the denominator of the fraction. Reduce the fraction if possible.

3. $6\% = \frac{6}{100} = \frac{3}{50}$

5. $33\% = \frac{33}{100}$

7. $55\% = \frac{55}{100} = \frac{11}{20}$

9. $75\% = \frac{75}{100} = \frac{3}{4}$

11. $20\% = \frac{20}{100} = \frac{1}{5}$

13. $9.5\% = 0.005 = \frac{95}{1000} = \frac{19}{200}$

15. $17.5\% = 0.175 = \frac{175}{1000} = \frac{7}{40}$

17. $64.8\% = 0.648 = \frac{648}{1000} = \frac{81}{125}$

19. $71.25\% = 0.7125 = \frac{7125}{10{,}000} = \frac{57}{80}$

21. $176\% = \frac{176}{100} = \frac{44}{25} = 1\frac{19}{25}$

23. $340\% = \frac{340}{100} = \frac{17}{5} = 3\frac{2}{5}$

25. $1200\% = \frac{1200}{200} = 12$

27. $2\frac{1}{6}\% = \frac{\frac{13}{6}}{100} = \frac{13}{600}$

29. $12\frac{1}{2}\% = \frac{\frac{25}{2}}{100} = \frac{25}{200} = \frac{1}{8}$

31. $8\frac{4}{5}\% = \frac{\frac{44}{5}}{100} = \frac{44}{500} = \frac{11}{125}$

33. $3.3\% = 0.033 = \frac{33}{1000}$

35. $2\frac{6}{11}\% = \frac{\frac{28}{11}}{100} = \frac{28}{1100} = \frac{7}{275}$

37. $\frac{3}{4} = 0.75 = 75\%$

39. $\frac{7}{10} = 0.7 = 70\%$

41. $\frac{7}{20} = 0.35 = 35\%$

43. $\frac{7}{25} = 0.28 = 28\%$

45. $\frac{11}{40} = 0.275 = 27.5\%$

47. $\frac{18}{5} = 3.6 = 360\%$

49. $2\frac{1}{2} = 2.5 = 250\%$

51. $4\frac{1}{8} = \frac{33}{8} = 4.125 = 412.5\%$

53. $\frac{1}{3} = 0.3333 = 33.33\%$

55. $\frac{3}{7} \approx 0.4286 = 42.86\%$

57. $\frac{17}{4} = 4.25 = 425\%$

59. $\frac{26}{50} = 0.52 = 52\%$

61.
```
      0.025
40)1.000
    80
    200
    200
      0
```
$\frac{1}{40} = 0.025 = 2.5\%$

63.
```
         0.01166
3000)35.00000
       3000
        5000
        3000
        20000
        18000
         20000
         18000
          2000
```
$\frac{35}{23000} \approx 0.0117 = 1.17\%$

65.
```
   0.37
8)3.00
   24
   60
   56
    4
```
$\frac{3}{8} = 0.37\frac{4}{8} = 0.37\frac{1}{2} = 37\frac{1}{2}\%$

67.
```
     0.07
40)3.00
    280
     20
```
$\frac{3}{40} = 0.07\frac{20}{40} = 0.07\frac{1}{2} = 7\frac{1}{2}\%$

69. $$\begin{array}{r} 0.83 \\ 6\overline{)5.00} \\ \underline{48} \\ 20 \\ \underline{18} \\ 2 \end{array}$$

$$\frac{5}{6} = 0.83\frac{2}{6} = 0.83\frac{1}{3} = 83\frac{1}{3}\%$$

71. $$\begin{array}{r} 0.22 \\ 9\overline{)2.00} \\ \underline{18} \\ 2 \end{array}$$

$$\frac{2}{9} = 0.22\frac{2}{9} = 22\frac{2}{9}\%$$

73. $$\begin{array}{r} 0.91666 \\ 12\overline{)11.00000} \\ \underline{108} \\ 20 \\ \underline{12} \\ 80 \\ \underline{72} \\ 80 \\ \underline{72} \\ 80 \\ \underline{72} \\ 8 \end{array}$$

$$\frac{11}{12} \approx 0.9167 = 91.67\%$$

$$\frac{11}{12};\ 0.9167;\ 67\%$$

75. $0.56 = 56\%$

$$0.56 = \frac{56}{100} = \frac{14}{25}$$

$$\frac{14}{25};\ 0.56;\ 56\%$$

77. $$0.005 = \frac{5}{1000} = \frac{1}{200}$$

$0.005 = 0.5\%$

79. $$\begin{array}{r} 0.555 \\ 9\overline{)\ 5.0} \\ \underline{45} \\ 50 \\ \underline{45} \\ 50 \\ \underline{45} \\ 5 \end{array}$$

$$\frac{5}{9} = 0.\overline{5}$$

$$\frac{5}{9};\ 0.5556;\ 55.56\%$$

81. $$\frac{1}{8} = 0.125$$

$$3\frac{1}{8}\% = 0.03125 \approx 0.0313$$

$$0.03125 = \frac{3125}{100{,}000} = \frac{1}{32}$$

$$\frac{1}{32};\ 0.0313;\ 3\frac{1}{8}\%$$

83. $$\begin{aligned} 28\frac{15}{16} &= \frac{28\frac{15}{16}}{100} \\ &= 28\frac{15}{16} \div 100 \\ &= \frac{463}{16} \times \frac{1}{100} \\ &= \frac{463}{1600} \end{aligned}$$

85. $$\begin{aligned} \frac{123}{800} &= \frac{n}{100} \\ 800 \times n &= 123 \times 100 \\ \frac{800 \times n}{800} &= \frac{12{,}300}{800} \\ n &= 15.375 \end{aligned}$$

$$\frac{123}{800} = 15.375\%$$

87. It will have at least two zeros to the right of the decimal point.

Cumulative Review

89. $\frac{15}{n} = \frac{8}{3}$

$15 \times 3 = n \times 8$

$\frac{45}{8} = \frac{n \times 8}{8}$

$n = 5.625$

91.
$$\begin{array}{r} 10{,}041 \\ 986 \\ 4{,}283 \\ +\ 533{,}855 \\ \hline 549{,}165 \end{array}$$
549,165 documents

5.3A Exercises

1. What is 20% of \$300?

3. 20 baskets out of 25 shots is what percent?

5. This type is called "a percent problem when we do not know the base." We can translate this into an equation

$108 = 18\% \times n$

$108 = 0.18n$

$\frac{108}{0.18} = \frac{0.18n}{0.18}$

$600 = n$

7. What is 5% of 90?

$n = 5\% \times 90$

9. 70% of what is 2?

$70\% \times n = 2$

11. 17 is what percent of 85?

$17 = n \times 85$

13. What is 20% of 140?

$n = 20\% \times 140$

$n = 0.20 \times 140$

$n = 28$

15. Find 40% of 140.

$n = 40\% \times 140$

$n = 0.4 \times 140$

$n = 56$

17. What is 5% of \$480?

$n = 5\% \times 480$

$n = 0.05 \times 480$

$n = 24$

\$24

19. 2% of what is 26?

$2\% \times n = 26$

$0.02 \times n = 26$

$\frac{0.02 \times n}{0.02} = \frac{26}{0.02}$

$n = 1300$

21. 52 is 4% of what?

$52 = 4\% \times n$

$52 = 0.04 \times n$

$\frac{52}{0.04} = \frac{0.04n}{0.04}$

$1300 = n$

23. 22% of what is \$33?

$22\% \times n = 33$

$0.22 \times n = 33$

$\frac{0.22 \times n}{0.22} = \frac{33}{0.22}$

$n = 150$

\$150

25. What percent of 200 is 168?

$n \times 200 = 168$

$\frac{n \times 200}{200} = \frac{168}{200}$

$n = 0.84$

$n = 84\%$

27. 56 is what percent of 200?

$56 = n \times 200$

$\frac{56}{200} = \frac{n \times 200}{200}$

$0.28 = n$

$28\% = n$

29. 78 is what percent of 120?

$78 = n \times 120$

$\frac{78}{120} = \frac{n \times 120}{120}$

$0.65 = n$

$65\% = n$

31. 20% of 155 is what?

$20\% \times 155 = n$

$0.20 \times 155 = n$

$31 = n$

33. 170% of what is 144.5?

$170\% \times n = 144.5$

$1.70 \times n = 144.5$

$\frac{1.70 \times n}{1.70} = \frac{144.5}{1.70}$

$n = 85$

35. 84 is what percent of 700?

$84 = n \times 700$

$\frac{84}{700} = \frac{n \times 700}{700}$

$0.12 = n$

$12\% = n$

37. Find 0.4% of 820.

$n = 0.4\% \times 820$

$n = 0.004 \times 820$

$n = 3.28$

39. What percent of 35 is 22.4?

$n \times 35 = 22.4$

$\frac{n \times 35}{35} = \frac{22.4}{35}$

$n = 0.64$

$n = 64\%$

41. 89 is 20% of what?

$89 = 20\% \times n$

$89 = 0.2 \times n$

$\frac{89}{0.2} = \frac{0.2 \times n}{0.2}$

$445 = n$

43. 8 is what percent of 1000?

$8 = n \times 1000$

$\frac{8}{1000} = \frac{n \times 1000}{1000}$

$0.008 = n$

$0.8\% = n$

45. What is 16.5% of 240?

$n = 16.5\% \times 240$

$n = 0.165 \times 240$

$n = 39.6$

47. 44 is what percent of 55?

$44 = n \times 55$

$\frac{44}{55} = \frac{n \times 55}{55}$

$0.8 = n$

$n = 80\%$

49. 68 is what percent of 80?

$68 = n \times 80$

$\frac{68}{80} = \frac{n \times 80}{80}$

$0.85 = n$

$85\% = n$

85% were acceptable.

51. 44% of 1260 is what?

$$44\% \times 1260 = n$$
$$0.44 \times 1260 = n$$
$$554.4 = n$$

554 students

53. 60% of what is 24?

$$60\% \times n = 24$$
$$0.60 \times n = 24$$
$$\frac{0.60 \times n}{0.60} = \frac{24}{0.60}$$
$$n = 40$$

40 years

55. Find 12% of 30% of \$1600.

$$n = 12\% \times 30\% \times 1600$$
$$n = 0.12 \times 0.30 \times 1600$$
$$n = 57.60$$

\$57.60

57.

$$L + 30\% \times L = 520$$
$$100\% \times L + 30\% L = 520$$
$$130\% \times L = 520$$
$$1.3 \times L = 520$$
$$\frac{1.3 \times L}{1.3} = \frac{520}{1.3}$$
$$L = 400 \text{ feet}$$
$$\text{and } W = 30\% \times 400$$
$$= 120 \text{ feet}$$

Cumulative Review

59.

$$\begin{array}{r} 1.36 \\ \times\ 1.8 \\ \hline 1088 \\ 126 \\ \hline 2.448 \end{array}$$

61.

$$\begin{array}{r} 2834. \\ 0.06\overline{)170.04} \\ \underline{12} \\ 50 \\ \underline{48} \\ 20 \\ \underline{18} \\ 24 \\ \underline{24} \\ 0 \end{array}$$

5.3B Exercises

	p	b	a
1. 75% of 660 is 495.	75	660	495
3. What is 22% of 60?	22	60	a
5. 49% of what is 2450?	49	b	2450
7. 30 is what percent of 50?	p	50	30

9. 40% of 70 is what?

$$\frac{a}{70} = \frac{40}{100}$$
$$100a = 70 \times 40$$
$$\frac{100a}{100} = \frac{2800}{100}$$
$$a = 28$$

11. Find 280% of 70.

$$\frac{a}{70} = \frac{280}{100}$$
$$100a = 70 \times 280$$
$$\frac{100a}{100} = \frac{19,600}{100}$$
$$a = 196$$

13. 0.7% of 8000 is what?

$$\frac{a}{8000} = \frac{0.7}{100}$$
$$100a = 8000 \times 0.7$$
$$\frac{100a}{100} = \frac{5600}{100}$$
$$a = 56$$

15. 20 is 25% of what?

$$\frac{20}{b}=\frac{25}{100}$$
$$20\times 100=25b$$
$$\frac{2000}{25}=\frac{25b}{25}$$
$$80=b$$

17. 250% of what is 200?

$$\frac{200}{b}=\frac{250}{100}$$
$$200\times 100=250b$$
$$\frac{20,000}{250}=\frac{250b}{250}$$
$$80=b$$

19. 3000 is 0.5% of what?

$$\frac{3000}{b}=\frac{0.5}{100}$$
$$3000\times 100=0.5b$$
$$\frac{300,000}{0.5}=\frac{0.5b}{0.5}$$
$$600,000=b$$

21. 56 is what percent of 280?

$$\frac{p}{100}=\frac{56}{280}$$
$$280p=56\times 100$$
$$\frac{280p}{280}=\frac{5600}{280}$$
$$p=20$$

20%

23. What percent of 260 is 10.4?

$$\frac{p}{100}=\frac{10.4}{260}$$
$$260p=100\times 10.4$$
$$\frac{260p}{260}=\frac{1040}{260}$$
$$p=4$$

4%

25. 25% of 88 is what?

$$\frac{25}{100}=\frac{a}{88}$$
$$25\times 88=100a$$
$$\frac{2200}{100}=\frac{100a}{100}$$
$$22=a$$

27. 300% of what is 120?

$$\frac{300}{100}=\frac{120}{b}$$
$$300b=100\times 120$$
$$\frac{300b}{300}=\frac{12,000}{300}$$
$$b=40$$

29. 82 is what percent of 500?

$$\frac{p}{100}=\frac{82}{500}$$
$$500p=100\times 82$$
$$\frac{500p}{500}=\frac{8200}{500}$$
$$p=16.4$$

16.4%

31. Find 0.7% of 520.

$$\frac{a}{520}=\frac{0.7}{100}$$
$$100a=520\times 0.7$$
$$\frac{100a}{100}=\frac{364}{100}$$
$$a=3.64$$

32. What percent of 66 is 16.5?

$$\frac{p}{100}=\frac{16.5}{66}$$
$$66p=100\times 16.5$$
$$\frac{66p}{66}=\frac{1650}{66}$$
$$p=25$$

25%

35. 68 is 40% of what?

$$\frac{40}{100} = \frac{68}{b}$$
$$40b = 100 \times 68$$
$$\frac{40b}{40} = \frac{6800}{40}$$
$$b = 170$$

37. 5% of what is $35?

$$\frac{5}{100} = \frac{35}{b}$$
$$5b = 100 \times 35$$
$$\frac{5b}{5} = \frac{3500}{5}$$
$$b = 700$$

$700

39. 3.90 is what percent of $26.00?

$$\frac{p}{100} = \frac{3.90}{26.00}$$
$$26.00p = 100 \times 3.90$$
$$\frac{26.00p}{26.00} = \frac{390}{26.00}$$
$$p = 15$$

15%

41. What is 15% of 120?

$$\frac{a}{120} = \frac{15}{100}$$
$$100a = 120 \times 15$$
$$\frac{100a}{100} = \frac{1800}{100}$$
$$a = 18$$

18 gallons

43. What is 24% of 9500?

$$n = 24\% \times 9500$$
$$n = 0.24 \times 9500$$
$$n = 2280$$

$2280

45.

$$\begin{array}{r} 2167 \\ 2963 \\ 956 \\ 621 \\ +\quad 74 \\ \hline 6781 \end{array}$$

What percent of 6781 is 2963?

$$n \times 6781 = 2963$$
$$\frac{n \times 6781}{6781} = \frac{2963}{6781}$$
$$n = 0.437$$
$$n = 43.7\%$$

47. Corrections: $n = 125\% \times 2963$

$$n = 1.25 \times 2963$$
$$n = 3703.75$$

Govt. & Admin.

$$n = 130\% \times 2167$$
$$n = 1.30 \times 2167$$
$$n = 2817.1$$

Total is:

$$\begin{array}{r} 3703.75 \\ 2817.10 \\ 956.00 \\ 621.00 \\ +\quad 74.00 \\ \hline 8171.85 \end{array}$$

What percent of 8171.85 is 621?

$$n \times 8171.85 = 621$$
$$\frac{n \times 8171.85}{8171.85} = \frac{621}{8171.85}$$
$$n = 0.076$$
$$n = 7.6\%$$

Cumulative Review

49. $\frac{4}{5}+\frac{8}{9}=\frac{4}{5}\times\frac{9}{9}+\frac{8}{9}\times\frac{5}{5}$

$=\frac{36}{45}+\frac{40}{45}$

$=\frac{76}{45}$

$=1\frac{31}{45}$

51. $\left(2\frac{4}{5}\right)\left(1\frac{1}{2}\right)=\frac{14}{5}\times\frac{3}{2}$

$=\frac{42}{10}=\frac{21}{5}$

$=4\frac{1}{5}$

How Am I Doing? Sections 5.1-5.3

1. $0.17 = 17\%$

2. $0.387 = 38.7\%$

3. $1.34 = 134\%$

4. $8.94 = 894\%$

5. $0.006 = 0.6\%$

6. $0.0004 = 0.04\%$

7. $\frac{17}{100}=17\%$

8. $\frac{89}{100}=89\%$

9. $\frac{13.4}{100}=13.4\%$

10. $\frac{19.8}{100}=19.8\%$

11. $\frac{6\frac{1}{2}}{100}=6\frac{1}{2}\%$

12. $\frac{1\frac{3}{8}}{100}=1\frac{3}{8}\%$

13. $10\overline{)8.0}$ gives 0.8

$\frac{8}{10}=0.8=80\%$

14. $40\overline{)1.000}$ gives $.025$

$$\begin{array}{r} 80 \\ \hline 200 \\ 200 \\ \hline 0 \end{array}$$

$\frac{1}{40}=0.025=2.5\%$

15. $20\overline{)52.0}$ gives 2.6

$$\begin{array}{r} 40 \\ \hline 120 \\ 120 \\ \hline 0 \end{array}$$

$\frac{52}{20}=2.6=260\%$

16. $16\overline{)17.0000}$ gives 1.0

$$\begin{array}{r} 16 \\ \hline 100 \\ 96 \\ \hline 40 \\ 32 \\ \hline 80 \\ 80 \\ \hline 0 \end{array}$$

$\frac{17}{16}=1.0625=106.25\%$

17.
$$\begin{array}{r} .71428 \\ 7\overline{)5.000} \\ \underline{49} \\ 10 \\ \underline{7} \\ 30 \\ \underline{28} \\ 20 \\ \underline{14} \\ 60 \\ \underline{56} \\ 4 \end{array}$$

$\frac{5}{7} \approx 0.7143 = 71.43\%$

18.
$$\begin{array}{r} .28571 \\ 7\overline{)2.00000} \\ \underline{14} \\ 60 \\ \underline{56} \\ 40 \\ \underline{35} \\ 50 \\ \underline{49} \\ 10 \end{array}$$

$\frac{2}{7} \approx 0.2857 = 28.57\%$

19.
$$\begin{array}{r} .95652 \\ 23\overline{)22.00000} \\ \underline{207} \\ 130 \\ \underline{115} \\ 150 \\ \underline{138} \\ 120 \\ \underline{115} \\ 50 \\ \underline{46} \\ 4 \end{array}$$

$\frac{22}{23} \approx 0.9565 = 95.65\%$

20.
$$\begin{array}{r} .68421 \\ 19\overline{)13.00000} \\ \underline{114} \\ 160 \\ \underline{152} \\ 80 \\ \underline{76} \\ 40 \\ \underline{38} \\ 20 \\ \underline{19} \\ 1 \end{array}$$

$\frac{13}{19} \approx 0.6842 = 68.42\%$

21. $4\frac{2}{5} = \frac{22}{5} = 4.4 = 440\%$

22. $2\frac{3}{4} = \frac{11}{4} = 2.75 = 275\%$

23. $\frac{1}{300} = \frac{1}{3}\left(\frac{1}{100}\right) = \frac{1}{3}\% - 0.33\%$

24. $\frac{1}{400} = \frac{1}{4}\left(\frac{1}{100}\right) = \frac{1}{4}\% = 0.25\%$

25. $22\% = \frac{22}{100} = \frac{11}{50}$

26. $53\% = \frac{53}{100}$

27. $150\% = \frac{150}{100} = \frac{3}{2}$ or $1\frac{1}{2}$

28. $160\% = \frac{160}{100} = \frac{8}{5}$ or $1\frac{3}{5}$

29. $6\frac{1}{3}\% = \frac{6\frac{1}{3}}{100} = \frac{\frac{19}{3}}{100} = \frac{19}{300}$

30. $4\frac{2}{3}\% = \frac{14}{3}\% = \frac{\frac{14}{3}}{100} = \frac{14}{300} = \frac{7}{150}$

31. $51\frac{1}{4}\% = \frac{\frac{205}{4}}{100} = \frac{205}{400} = \frac{41}{80}$

32. $43\frac{3}{4}\% = \frac{\frac{175}{4}}{100} = \frac{175}{400} = \frac{7}{16}$

33. Find 24% of 230.

$n = 24\% \times 230$

$n = 0.24(230)$

$n = 55.2$

34. What is 78% of 62?

$n = 78\% \times 62$

$n = 0.78(62)$

$n = 48.36$

35. 68 is what percent of 72?

$\frac{p}{100} = \frac{68}{72}$

$72p = 68 \times 100$

$\frac{72p}{72} = \frac{6800}{72}$

$p = 94.44$

94.44%

36. What percent of 76 is 34?

$\frac{p}{100} = \frac{34}{76}$

$76p = 34 \times 100$

$\frac{76p}{76} = \frac{3400}{76}$

$p = 44.74$

44.74%

37. 8% of what number is 240?

$\frac{8}{100} = \frac{240}{b}$

$8b = 240 \times 100$

$\frac{8b}{8} = \frac{24,000}{8}$

$b = 3000$

38. 354 is 40% of what number?

$\frac{40}{100} = \frac{354}{b}$

$40b = 354 \times 100$

$\frac{40b}{40} = \frac{35,400}{40}$

$b = 885$

5.4 Exercises

1. $n \times 2.5\% = 4500$

$\frac{n \times 0.025}{0.025} = \frac{4500}{0.025}$

$n = 180,000$

180,000 pencils

3. $80.50 = 115\% \times n$

$80.5 = 1.15n$

$\frac{80.5}{1.15} = \frac{1.15n}{1.15}$

$\$70 = n$

5. $n \times 6000 = 431$

$\frac{6000n}{6000} = \frac{431}{6000}$

$n = 0.0718$

$n = 7.18\%$

7. $n = 6\% \times 65$

$n = 0.06 \times 65$

$n = 3.9$

$3.90

9. $n \times 4\% = 9.60$

$0.04n = 9.6$

$\frac{0.04n}{0.04} = \frac{9.6}{0.04}$

$n = 240$

$240

11. $n \times 5060 = 1265$

$$\frac{n \times 5060}{5060} = \frac{1265}{5060}$$
$$n = 0.25$$

25%

13. $n \times 75\% = 7,200,000$

$$\frac{0.75n}{0.75} = \frac{7,200,000}{0.75}$$
$$n = 9,600,000$$

\$9,600,000

15. $n \times 10,001 = 248$

$$\frac{10,001n}{10,001} = \frac{248}{10,001}$$
$$n = 0.0248$$

2.48%

17. $n = 0.9\% \times 24,000$

$n = 0.009 \times 24,000$

$n = 216$

216 babies

19. $105\% \times n = 800$

$$\frac{1.05n}{1.05} = \frac{800}{1.05}$$
$$n = 761.90$$

\$761.90

21. $100\% \times n + 9\% \times n = 163,500$

$109\% \times n = 163,500$

$1.09n = 163,500$

$$\frac{1.09n}{1.09} = \frac{163,500}{1.09}$$
$$n = 150,000$$

\$150,000

23. $3\% + 8\% = 11\%$

$n = 11\% \times 20,000$

$n = 0.11 \times 20,000$

$n = 2200$

2200 pounds

25. a. $15\% + 12\% + 10\% = 37\%$

$n = 37\% \times 33,000,000$

$n = 0.37 \times 33,000,000$

$n = 12,210,000$

\$12,210,000 for personnel, food, & decorations

b.
$$\begin{array}{r} 33,000,000 \\ -\ 12,210,000 \\ \hline 20,790,000 \end{array}$$

\$20,790,000 for security, facility rental, & all other expenses

27. $n = 35\% \times 190$

$n = 0.35 \times 190$

$n = 66.6$

Discount is \$65.50

Sales price is \$190 – \$65.50 = \$123.50

29. $n = 15\% \times 400$

$n = 0.15 \times 400$

$n = 60$

Save \$60 with coupon book.

Total savings \$60 – \$30 = \$30.

31. a. $n = 8\% \times 16,000$

$n = 0.08 \times 16,000$

$n = 1280$

Discount is \$1280.

b. Cost of motorcycle is

\$16,000 – \$1280 = \$14,720

33. a. $n = 33\% \times 1110$

$n = 0.33 \times 1110$

$n = 333$

Discount is \$333.

b. Cost of clock is

\$1110 – \$333 = \$777

Cumulative Review

35. 1,698,481 rounds to 1,698,000

37. 1.63474 rounds to 1.63

39. 0.055613 rounds to 0.0556

5.5 Exercises

1. $\begin{aligned}\text{Commission} &= 2\% \times 170{,}000\\ &= 0.02\times 170{,}000\\ &= 3400\end{aligned}$

$3400

3. $\begin{aligned}\text{Total income} &= 300 + 4\% \times 96{,}000\\ &= 300 + 0.04\times 96{,}000\\ &= 300 + 3840\\ &= 4140\end{aligned}$

$4140

5. Increase $= 330 - 275 = 55$

$n \times 275 = 55$

$\dfrac{n\times 275}{275} = \dfrac{55}{275}$

$n = 0.2$

20% increase

7. Decrease $= 2000 - 1200 = 800$

$n \times 2000 = 800$

$\dfrac{n\times 2000}{2000} = \dfrac{800}{2000}$

$n = 0.4$

40% decrease

9. $\begin{aligned}I &= P\times R\times T\\ &= 2000\times 7\%\times 1\\ &= 2000\times 0.07\times 1\\ &= 140\end{aligned}$

$140

11. $\begin{aligned}I &= P\times R\times T\\ &= 500\times 1.5\%\times 1\\ &= 500\times 0.015\times 1\\ &= 7.5\end{aligned}$

$7.50

13. 3 months $= \dfrac{3}{12} = \dfrac{1}{4}$ year

$\begin{aligned}I &= P\times R\times T\\ &= 12{,}000\times 16\%\times \frac{1}{4}\\ &= 12{,}000\times 0.16\times \frac{1}{4}\\ &= 1920\times \frac{1}{4}\\ &= 480\end{aligned}$

$480

15. $\begin{aligned}\text{Rate} &= \frac{72{,}000}{12{,}000{,}000}\\ &= 0.006\\ &= 0.6\%\end{aligned}$

0.6%

17. $\begin{aligned}\text{Sales total} &= \frac{48{,}000}{3\%}\\ &= \frac{48{,}000}{0.03}\\ &= 1{,}600{,}000\end{aligned}$

$1,600,000

19. $\begin{aligned}\text{Spending} &= 15\%\times 265\\ &= 0.15\times 265\\ &= 39.75\end{aligned}$

$39.75

21. $\begin{aligned}\text{Thin mints} &= 25\%\times 156\\ &= 0.25\times 156\\ &= 39\end{aligned}$

39 boxes

23. $\text{Decrease} = 17.32 - 14.61$
$= 2.71$
$11 \times 17.32 = 2.71$
$\frac{n \times 17.32}{17.32} = \frac{2.71}{17.32}$
$n = 0.156$
$\approx 16\%$

25. $\text{Increase} = 72 - 40 = 32$
$n \times 40 = 32$
$\frac{n \times 40}{40} = \frac{32}{40}$
$n = 0.8$
80% increase

27. a. $n = 2.3\% \times 3700$
$n = 0.023 \times 3700$
$n = 85.1$
$85.10

b.
$$\begin{array}{r} 3700 \\ +\ \ 85.10 \\ \hline \$3785.10 \end{array}$$

29. a. $\text{Purchases} = 52 + 38 + 26$
$= 116$
$\text{Tax} = 6\% \times 116$
$= 0.06 \times 116$
$= \$6.96$

b. $\text{Total cost} = \$116 + \$6.96$
$= \$122.96$

31. $n = 114\% \times 9500$
$n = 1.14 \times 9500$
$n = 10,830$
$10,830

33. a. $n = 8\% \times 349,000$
$n = 0.08 \times 349,000$
$n = 27,920$
$27,920

b.
$$\begin{array}{r} 349,000 \\ +\ \ 27,920 \\ \hline \$321,080 \end{array}$$

35. $n \times 840 = 814$
$\frac{n \times 840}{840} = \frac{814}{840}$
$n = 0.969$
96.9%

37. $\text{Tax} = 4.6\% \times 18,456.82$
$= 0.046 \times 18,456.82$
≈ 849.01
$849.01

Cumulative Review

39. $3(12-6) - 4(12 \div 3) = 3(6) - 4(4)$
$= 18 - 16$
$= 2$

41.
$$\left(\frac{5}{2}\right)\left(\frac{1}{3}\right) - \left(\frac{2}{3} - \frac{1}{3}\right)^2 = \left(\frac{5}{2}\right)\left(\frac{1}{3}\right) - \left(\frac{1}{3}\right)^2$$
$$= \left(\frac{5}{2}\right)\left(\frac{1}{3}\right) - \frac{1}{9}$$
$$= \frac{5}{6} - \frac{1}{9}$$
$$= \frac{15}{18} - \frac{2}{18}$$
$$= \frac{13}{18}$$

Putting Your Skills To Work

1. $4,058,000 - 1,810,000$
$= 2,248,000$ students

2. $1,075,000 - 858,000$
$= 217,000$ students

3. a.

$$\begin{array}{r} 283,000 \\ 1,075,000 \\ +\ 4,058,000 \\ \hline 5,416,000 \end{array}$$

$$n \times 5,416,000 = 1,075,000$$

$$n = \frac{1,075,000}{5,416,000}$$

$$n = 0.198$$

19.8%

b.

$$\begin{array}{r} 411,000 \\ 1,230,000 \\ +\ 1,810,000 \\ \hline 3,451,000 \end{array}$$

$$n \times 3,451,000 = 1,230,000$$

$$n = \frac{1,230,000}{3,451,000}$$

$$n = 0.356$$

35.6% studied French in 1970. The percentage is less in 2000.

4. a. $n \times 5,416,,000$

$$n = \frac{283,000}{5,416,000}$$

$$n = 0.052$$

5.2%

b. $n \times 3,451,000 = 411,000$

$$n = \frac{411,000}{3,451,000}$$

$$n = 0.119$$

11.9% studied German in 1970.
The percentage is less in 2000.

5. $n = 24\%$ of 1,075,000

$$n = 0.24 \times 1,075,000$$

$$n = 258,000$$

258,000 more students

6. $n = 6\%$ of 4,058,000

$$n = 0.06 \times 4,058,000$$

$$n = 243,480$$

243,480 more students

Chapter 5 Review Problems

1. $0.62 = 62\%$

2. $0.43 = 43\%$

3. $0.372 = 37.2\%$

4. $0.529 = 52.9\%$

5. $1.05 = 105\%$

6. $2.1 = 210\%$

7. $2.52 = 252\%$

8. $4.37 = 437\%$

9. $1.036 = 103.6\%$

10. $1.052 = 105.2\%$

11. $0.006 = 0.6\%$

12. $0.002 = 0.2\%$

13. $\frac{12.5}{100} = 12.5\%$

14. $\frac{8.3}{100} = 8.3\%$

15. $\frac{4\frac{1}{12}}{100} = 4\frac{1}{12}\%$

16. $\frac{3\frac{5}{12}}{100} = 3\frac{5}{12}\%$

17. $\frac{317}{100} = 317\%$

18. $\frac{225}{100} = 225\%$

19. $\frac{19}{25} = 0.76 = 76\%$

20. $\frac{13}{25} = 0.52 = 52\%$

21. $\frac{11}{20} = 0.55 = 55\%$

22. $\frac{9}{40} = 0.225 = 22.5\%$

23. $\frac{5}{11} \approx 0.4545 = 45.45\%$

24. $\frac{4}{9} \approx 0.4444 = 44.44\%$

25. $2\frac{1}{4} = 2.25 = 225\%$

26. $3\frac{3}{4} = 3.75 = 375\%$

27. $2\frac{7}{9} \approx 2..7778 = 277.78\%$

28. $5\frac{5}{9} \approx 5.5556 = 555.56\%$

29. $\frac{152}{80} = 1.9 = 190\%$

30. $\frac{200}{80} = 2.5 = 250\%$

31. $\frac{3}{800} \approx 0.0038 = 0.38\%$

32. $\frac{5}{800} \approx 0.0063 = 0.63\%$

33. $32\% = 0.32$

34. $68\% = 0.68$

35. $82.7\% = 0.827$

36. $59.6\% = 0.596$

37. $236\% = 2.36$

38. $177\% = 1.77$

39. $32\frac{1}{8} = 32.125\%$
$= 0.32125$

40. $26\frac{3}{8}\% = 26.375\%$
$= 0.26375$

41. $72\% = \frac{72}{100} = \frac{72 \div 4}{100 \div 4}$
$= \frac{18}{25}$

42. $92\% = \frac{92}{100} = \frac{92 \div 4}{100 \div 4}$
$= \frac{23}{25}$

43. $185\% = \frac{185}{100} = \frac{185 \div 5}{100 \div 5}$
$= \frac{37}{20}$

44. $225\% = \frac{225}{100} = \frac{225 \div 25}{100 \div 25}$
$= \frac{9}{4}$

45. $16.4\% = 0.164$
$= \frac{164}{1000} = \frac{164 \div 4}{1000 \div 4}$
$= \frac{41}{250}$

46. $30.5\% = 0.305 = \frac{305}{1000}$
$= \frac{305 \div 5}{1000 \div 5}$
$= \frac{61}{200}$

47. $31\frac{1}{4}\% = \frac{31\frac{1}{4}}{100} = 31\frac{1}{4} \div 100$
$= \frac{125}{4} \times \frac{1}{100} = \frac{125}{400}$
$= \frac{125 \div 25}{400 \div 25}$
$= \frac{5}{16}$

48. $43\frac{3}{4}\% = \frac{43\frac{3}{4}}{100}$
$= 43\frac{3}{4} \div 100$
$= \frac{175}{4} \times \frac{1}{100}$
$= \frac{175}{400} = \frac{175 \div 25}{400 \div 25}$
$= \frac{7}{16}$

49. $0.05\% = 0.0005$
$= \frac{5}{10,000}$
$= \frac{5 \div 5}{10,000 \div 5}$
$= \frac{1}{2000}$

50. $0.06\% = 0.0006$
$= \frac{6}{10,000} = \frac{6 \div 2}{10,000 \div 2}$
$= \frac{3}{5000}$

51. $5\overline{)3.0}$ quotient 0.6
$\underline{30}$
0
$\frac{3}{5}$; 0.6; 60%

52. $\frac{7}{10} = 0.7 = 70\%$
$\frac{7}{10}$; 0.7; 70%

53. $37.5\% = 0.375$
$= \frac{375}{1000} = \frac{3}{8}$
$\frac{3}{8}$; 0.375; 37.5%

54. $56.25\% = 0.5625$
$= \frac{5625}{10,000} = \frac{9}{16}$
$\frac{9}{16}$; 0.5625; 56.25%

55. $0.008 = 0.8\%$
$0.008 = \frac{8}{1000} = \frac{1}{125}$
$\frac{1}{125}$; 0.008; 0.8%

56. $0.45 = 45\%$
$0.45 = \frac{45}{100} = \frac{9}{20}$
$\frac{9}{20}$; 0.45; 45%

57. What is 20% of 85?

$n = 20\% \times 85$

$n = 0.2 \times 85$

$n = 17$

58. What is 25% of 92?

$n = 25\% \times 92$

$n = 0.25 \times 92$

$n = 23$

59. 18 is 20% of what number?

$20\% \times n = 18$

$0.2n = 18$

$\frac{0.2n}{0.2} = \frac{18}{0.2}$

$n = 90$

60. 70 is 40% of what number?

$40\% \times n = 70$

$0.4n = 70$

$\frac{0.4n}{0.4} = \frac{70}{0.4}$

$n = 175$

61. 50 is what percent of 130?

$50 = n \times 130$

$\frac{50}{130} = \frac{n \times 130}{130}$

$0.3846 = n$

38.46%

62. 70 is what percent of 180?

$70 = n \times 180$

$\frac{70}{1380} = \frac{n \times 1380}{180}$

$0.3889 = n$

38.89%

63. Find 162% of 60.

$n = 162\% \times 60$

$n = 162 \times 60$

$n = 97.2$

64. Find 124% of 80.

$n = 124\% \times 80$

$n = 1.24 \times 80$

$n = 99.2$

65. 92% of what number is 147.2?

$92\% \times n = 147.2$

$0.92n = 147.2$

$\frac{0.92n}{0.92} = \frac{147.2}{0.92}$

$n = 160$

66. 68% of what number is 95.2?

$68\% \times n = 95.2$

$0.68n = 95.2$

$\frac{0.68n}{0.68} = \frac{95.2}{0.68}$

$n = 140$

67. What percent of 70 is 14?

$n \times 70 = 14$

$\frac{n \times 70}{70} = \frac{14}{70}$

$n = 0.2$

20%

68. What percent of 60 is 6?

$n \times 60 = 6$

$\frac{n \times 60}{60} = \frac{6}{60}$

$n = 0.1$

10%

69. $n = 34\% \times 150$

$n = 0.34 \times 150$

$n = 51$

51 students

70. $n = 64\% \times 150$

$n = 0.64 \times 150$

$n = 96$

96 trucks

71. $n \times 61\% = 6832$

$0.61n = 6832$

$\frac{0.61n}{0.61} = \frac{6832}{0.61}$

$n = 11,200$

\$11,200

72. $n \times 12\% = 9624$

$0.12n = 9624$

$\frac{0.12n}{0.12} = \frac{9624}{0.12}$

$n = 80,200$

\$80,200

73. Days $= 29 + 31 + 30 = 90$

Rain days $= 20 + 18 + 16 = 54$

$90 \times n = 54$

$\frac{90 \times n}{90} = \frac{54}{90}$

$n = 0.6$

60%

74. $600 \times n = 45$

$\frac{600 \times n}{600} = \frac{45}{600}$

$n = 0.075$

7.5%

75. What is 5% of 3670?

$n = 5\% \times 3670$

$n = 0.05 \times 3670$

$n = 183.5$

\$183.50

76. What is 6% of 12,600?

$n = 6\% \times 12,600$

$n = 0.06 \times 12,600$

$n = 756$

\$756

77. $38\% \times n = 684$

$0.38n = 684$

$\frac{0.38n}{0.38} = \frac{684}{0.38}$

$n = 1800$

\$1800

78. $\text{Rate} = \frac{26,000}{650,500}$

$= 0.04 = 4\%$

79. $\text{Rate} = \frac{5010}{83,500}$

$= 0.06 = 6\%$

80. Commission $= 7.5\% \times 16,000$

$= 0.075 \times 16,000$

$= 1200$

\$1200

81. a. Discount $= 20\% \times 1595$

$= 0.2 \times 1595$

$= 319$

\$319

b. Total pay $= 1595 - 319$

$= 1276$

\$1276

82. a. Rebate $= 12\% \times 2125$

$= 0.12 \times 2125$

$= 255$

\$255

b. Computer costs $= \$2125 - \$255$

$= \$1870$

83. Increase = 4618 − 3243 = 1375

$$\frac{p}{100} = \frac{1375}{3243}$$
$$3243p = 100 \times 1375$$
$$\frac{3243p}{3243} = \frac{137,500}{3243}$$
$$p = 42.40$$

42.40%

84. Increase = 360.10 − 313.80 = 46.3

$$\frac{p}{100} = \frac{46.3}{313.80}$$
$$313.80p = 100 \times 46.3$$
$$\frac{313.80p}{313.80} = \frac{4630}{313.80}$$
$$p = 14.75$$

14.75%

85. a. Discount = 14% of 24,000
$$= 0.14 \times 24,000$$
$$= \$3360$$

b. Cost = 24,000 − 3360
$$= \$20,640$$

86. $I = PRT$

a. $I = 6000(0.11)(0.5)$
$$= \$330$$

b. $I = 6000(0.11)(2)$
$$= \$1320$$

87. a. 3 months $= \frac{3}{12} = \frac{1}{4}$ year

$$I = P \times R \times T$$
$$= 3000 \times 8\% \times \frac{1}{4}$$
$$= 3000 \times 0.08 \times \frac{1}{4}$$
$$= 240 \times \frac{1}{4}$$
$$= 60$$

\$60

b. $I = P \times R \times T$
$$= 3000 \times 8\% \times 3$$
$$= 3000 \times 0.08 \times 3$$
$$= 240 \times 3$$
$$= 720$$

\$720

How Am I Doing? Chapter 5 Test

1. 0.57 = 57%

2. 0.01 = 1%

3. 0.008 = 0.8%

4. 12.8 = 1280%

5. 3.56 = 356%

6. $\frac{71}{100} = 71\%$

7. $\frac{1.8}{100} = 1.8\%$

8. $\frac{3\frac{1}{7}}{100} = 3\frac{1}{7}\%$

9. $$\begin{array}{r} 0.475 \\ 40\overline{)19.000} \\ \underline{160} \\ 300 \\ \underline{280} \\ 200 \\ \underline{200} \\ 0 \end{array}$$

$$\frac{19}{40} = 0.475 = 47.5\%$$

10. $$\begin{array}{r} 0.75 \\ 36\overline{)27.00} \\ \underline{252} \\ 180 \\ \underline{180} \\ 0 \end{array}$$

$$\frac{27}{36} = 75\%$$

11. $$\begin{array}{r} 3 \\ 75\overline{)225} \\ \underline{225} \\ 0 \end{array}$$

$$\frac{225}{75} = 3 = 300\%$$

12. $$\begin{array}{r} 0.75 \\ 4\overline{)3.00} \\ \underline{28} \\ 20 \\ \underline{20} \\ 0 \end{array}$$

$$1\frac{3}{4} = 1.75 = 175\%$$

13. $0.0825 = 8.25\%$

14. $3.024 = 302.4\%$

15. $152\% = \frac{152}{100} = \frac{38}{25} = 1\frac{13}{25}$

16. $$\begin{aligned} 7\frac{3}{4}\% &= \frac{7\frac{3}{4}}{100} \\ &= 7\frac{3}{4} \div 100 \\ &= \frac{31}{4} \times \frac{1}{100} \\ &= \frac{31}{400} \end{aligned}$$

17. $$\begin{aligned} n &= 40\% \times 50 \\ n &= 0.4 \times 50 \\ n &= 20 \end{aligned}$$

18. $$\begin{aligned} 33.8 &= 26\% \times n \\ 33.8 &= 0.26n \\ \frac{33.8}{0.26} &= \frac{0.26n}{0.26} \\ 130 &= n \end{aligned}$$

19. $$\begin{aligned} n \times 72 &= 40 \\ \frac{n \times 72}{72} &= \frac{40}{72} \\ n &\approx 0.5556 = 55.56\% \end{aligned}$$

20. $$\begin{aligned} n &= 0.8\% \times 25{,}000 \\ n &= 0.008 \times 25{,}000 \\ n &= 200 \end{aligned}$$

21. $$\begin{aligned} 16\% \times n &= 800 \\ 0.16n &= 800 \\ \frac{0.16n}{0.16} &= \frac{800}{0.16} \\ n &= 5000 \end{aligned}$$

22. $$\begin{aligned} 92 &= n \times 200 \\ \frac{92}{200} &= \frac{n \times 200}{200} \\ n &= 0.46 \\ n &= 46\% \end{aligned}$$

23. $$\begin{aligned} 132\% \times 530 &= n \\ 1.32 \times 530 &= n \\ 699.6 &= n \end{aligned}$$

24. $p \times 75 = 15$

$$\frac{75p}{75} = \frac{15}{75}$$

$$p = 0.2$$
$$= 20\%$$

20%

25. $n = 4\% \times 152{,}300$

$n = 0.04 \times 152{,}300$

$n = 6092$

\$6092

26. a. $33\% \times 457 = n$

$0.33 \times 457 = n$

$150.81 = n$

\$150.81

b. $457 - 150.81 = 306.19$

\$306.19

27. $75 = n \times 84$

$$\frac{75}{84} = \frac{n \times 84}{84}$$

$n \approx 0.8929 = 89.29\%$

28. Increase $= 228 - 185 = 43$

$$\frac{p}{100} = \frac{43}{185}$$

$185p = 100 \times 43$

$$\frac{185p}{185} = \frac{4300}{185}$$

$p = 23.24$

23.24%

29. $5160 = 43\% \times n$

$5160 = 0.43n$

$$\frac{5160}{0.43} = \frac{0.43n}{0.43}$$

$12{,}000 = n$

12,000 registered voters

30. a. $I = P \times R \times T$

$$= 3000 \times 0.16 \times \frac{6}{12}$$
$$= 240$$

\$240

b. $I = P \times R \times T$

$= 3000 \times 0.16 \times 2$

$= 960$

\$960

Cumulative Test for Chapters 1-5

1.
$$\begin{array}{r} 38 \\ 196 \\ +2007 \\ \hline 2241 \end{array}$$

2.
$$\begin{array}{r} 23{,}007 \\ -\ 14{,}563 \\ \hline 8{,}444 \end{array}$$

3.
$$\begin{array}{r} 126 \\ \times\ 42 \\ \hline 252 \\ 504\ \\ \hline 5292 \end{array}$$

4.
$$\begin{array}{r} 89 \\ 36\overline{)3204} \\ \underline{288}\ \\ 324 \\ \underline{324} \\ 0 \end{array}$$

5.
$$\begin{array}{r} 2\frac{1}{4} \\ +\ 3\frac{1}{3} \\ \hline \end{array} \qquad \begin{array}{r} 2\frac{3}{12} \\ +\ 3\frac{4}{12} \\ \hline 5\frac{7}{12} \end{array}$$

6.
$$\begin{array}{r} 5\frac{2}{5} \\ -\ 2\frac{7}{10} \\ \hline \end{array} \qquad \begin{array}{r} 5\frac{4}{10} \\ -\ 2\frac{7}{10} \\ \hline \end{array} \qquad \begin{array}{r} 4\frac{14}{10} \\ -\ 2\frac{7}{10} \\ \hline 2\frac{7}{10} \end{array}$$

7. $3\frac{1}{8}\times\frac{12}{5}=\frac{25}{8}\times\frac{12}{5}=\frac{5\times3}{2}$

$=\frac{15}{2}$ or $7\frac{1}{2}$

8. $\frac{5}{12}\div1\frac{3}{4}=\frac{5}{12}\div\frac{7}{4}$

$=\frac{5}{12}\times\frac{4}{7}$

$=\frac{5}{21}$

9. 77.18<u>3</u>2 rounds to 77.183

10.

$$\begin{array}{r} 5.600 \\ 3.210 \\ 18.300 \\ +\ \ 7.008 \\ \hline 34.118 \end{array}$$

11.

$$\begin{array}{r} 5.62 \\ \times\ \ 0.3 \\ \hline 1.686 \end{array}$$

12.

$$\begin{array}{r} 0.368 \\ 1.4\overline{)0.5152} \\ \underline{42} \\ 95 \\ \underline{84} \\ 112 \\ \underline{112} \\ 0 \end{array}$$

13. $\frac{36\text{ tiles}}{9\text{ sq. ft.}}=\frac{9\times4}{9}=4$ tiles/square foot

14. $\frac{20}{25}\stackrel{?}{=}\frac{300}{375}$

$20\times375\stackrel{?}{=}25\times300$

$7500=7500$ True

15. $\frac{8}{2.5}=\frac{n}{7.5}$

$8\times75=2.5\times n$

$\frac{60}{2.5}=\frac{2.5\times n}{2.5}$

$n=24$

16. $\frac{3\text{ faculty}}{19\text{ students}}=\frac{n\text{ faculty}}{4263\text{ students}}$

$3\times4263=19\times n$

$12,789=19\times n$

$\frac{12,789}{19}=\frac{19\times n}{19}$

$673\approx n$

$673=$ faculty

17. $0.023=2.3\%$

18. $\frac{46.8}{100}=46.8\%$

19. $1.98=198\%$

20.

$$\begin{array}{r} 0.0375 \\ 80\overline{)3.0000} \\ \underline{240} \\ 600 \\ \underline{560} \\ 400 \\ \underline{400} \\ 0 \end{array}$$

$\frac{3}{80}=0.0375=3.75\%$

21. $243\%=2.43$

22. $\begin{array}{r} 0.75 \\ 4\overline{)3.00} \\ \underline{20} \\ 20 \\ \underline{20} \\ 0 \end{array}$

$6\frac{3}{4}\% = 6.75\% = 0.0675$

23. What percent of 214 is 38?

$n \times 214 = 38$

$\frac{n \times 214}{214} = \frac{38}{214}$

$n \approx 0.1776$

$n = 17.76\%$

24. $n = 1.7\% \times 6740$

$n = 0.017 \times 6740$

$n = 114.58$

25. 219 is 73% of what number?

$219 = 73\% \times n$

$\frac{219}{0.73} = \frac{0.73 \times n}{0.73}$

$n = 300$

26. 95% of 200 is what number?

$n = 95\% \times 200$

$n = 0.95 \times 200$

$n = 190$

27. Decrease $= 20\% \times 680$

$= 0.2 \times 680$

$= 136$

Price $= 680 - 136 = 544$

\$544

28. $28\% \times n = 896$

$0.28n = 896$

$\frac{0.28n}{0.28} = \frac{896}{0.28}$

$n = 3200$

3200 students

29. $8.86 - 7.96 = 0.9$

$7.96 \times n = 0.9$

$\frac{7.96 \times n}{7.96} = \frac{0.9}{7.96}$

$n \approx 0.1131$

$n = 11.31\%$ increase

30. $I = P \times R \times T$

$= 1600 \times 0.11 \times 2$

$= 352$

\$352

Chapter 6

6.1 Exercises

1. We know that each mile has 5280 feet.. Each foot has 12 inches. Therefore, we know that one mile is $5280 \times 12 = 63{,}360$ inches. The unit fraction is $\dfrac{63{,}360 \text{ inches}}{1 \text{ mile}}$. So we multiply $23 \text{ miles} \times \dfrac{63{,}360 \text{ inches}}{1 \text{ mile}}$. The mile units divide out. We obtain 1,457,280 inches. Thus, $23 \text{ miles} = 1{,}457{,}280$ inches.

3. $1760 \text{ yards} = 1 \text{ mile}$

5. $1 \text{ ton} = 2000 \text{ pounds}$

7. $4 \text{ quarts} = 1 \text{ gallon}$

9. $1 \text{ quart} = 2 \text{ pints}$

11. $21 \text{ feet} \times \dfrac{1 \text{ yard}}{3 \text{ feet}} = 7 \text{ yards}$

13. $108 \text{ inches} \times \dfrac{1 \text{ foot}}{12 \text{ inches}} = 9 \text{ feet}$

15. $12 \text{ feet} \times \dfrac{12 \text{ inches}}{1 \text{ foot}} = 144 \text{ inches}$

17. $10{,}560 \text{ feet} \times \dfrac{1 \text{ mile}}{5280 \text{ feet}} = 2 \text{ miles}$

19. $7 \text{ miles} \times \dfrac{1760 \text{ yards}}{1 \text{ mile}} = 12{,}320 \text{ yards}$

21. $7 \text{ gallons} \times \dfrac{4 \text{ quarts}}{1 \text{ gallon}} = 28 \text{ quarts}$

23. $48\text{quarts} \times \dfrac{1 \text{ gallon}}{4 \text{ quarts}} = 12 \text{ gallons}$

25. $16 \text{ cups} \times \dfrac{8 \text{ fluid ounces}}{1 \text{ cup}} = 128 \text{ fluid ounces}$

27. $8\frac{1}{2} = \dfrac{17}{2} \text{ gallons} \times \dfrac{8 \text{ pints}}{\text{gallon}} = 68 \text{ pints}$

29. $12 \text{ weeks} \times \dfrac{7 \text{ days}}{1 \text{ week}} = 84 \text{ days}$

31. $960 \text{ seconds} \times \dfrac{1 \text{ minute}}{60 \text{ seconds}} = 16 \text{ minutes}$

33. $8 \text{ ounces} \times \dfrac{1 \text{ pound}}{16 \text{ ounces}} = 0.5 \text{ pound}$

35. $12{,}500 \text{ pounds} \times \dfrac{1 \text{ ton}}{2000 \text{ pounds}} = 6.25 \text{ tons}$

37. $15 \text{ pints} \times \dfrac{2 \text{ cup}}{1\text{pint}} = 30 \text{ cups}$

39. $2.25 \text{ pounds} \times \dfrac{16 \text{ ounces}}{1 \text{ pound}} = 36 \text{ ounces}$

41. $75 \text{ inches} \times \dfrac{1 \text{ foot}}{12 \text{ inches}} = 6.25 \text{ feet}$

43. $26.2 \text{ miles} \times \dfrac{5280 \text{ feet}}{1 \text{ mile}} = 138{,}336 \text{ feet}$

45. $19{,}336 \text{ feet} \times \dfrac{1 \text{ mile}}{5280 \text{ feet}} = 3.66 \text{ miles}$

47. $26 \text{ ounces} \times \dfrac{1 \text{ pound}}{16 \text{ ounces}} \times \dfrac{\$6.00}{1 \text{ pound}} = \$9.75$

49. a. $2 \text{ feet} \times \dfrac{12 \text{ inches}}{1 \text{ foot}} = 24 \text{ inches}$

$2 \text{ feet } 3 \text{ inches} = 24 + 3 = 27 \text{ inches}$

$3 \text{ feet} \times \dfrac{12 \text{ inches}}{1 \text{ foot}} = 36 \text{ inches}$

$3 \text{ feet } 9 \text{ inches} = 36 + 9 = 45 \text{ inches}$

$$\text{perimeter} = 2(27) + 2(45) = 54 + 90 = 144 \text{ inches}$$

b. $\text{cost} = \$0.85 \times 144 = \$122.40$

51. $7200 \text{ quarts} \times \frac{4 \text{ cups}}{1 \text{ quart}} = 28{,}800 \text{ cups}$

53. 1760 rounds to 2000

$6 \text{ miles} \times \frac{2000 \text{ yards}}{1 \text{ mile}} \approx 12{,}000 \text{ yards}$

55. 33,000 rounds to 30,000

5280 rounds to 5000

$30{,}000 \text{ feet} \times \frac{1 \text{ mile}}{5000 \text{ feet}} \approx 6 \text{ miles}$

57. $12{,}800 \text{ nautical miles} \times \frac{38 \text{ land miles}}{33 \text{ nautical miles}}$

$\approx 14{,}739 \text{ land miles}$

Cumulative Review

59. $560 - 515 = \$45$ per month

Savings $= 45 \times 20 \times 12 = \$10{,}800$

61. $\frac{n}{115} = \frac{7}{5}$

$n \times 5 = 7 \times 115$

$\frac{n \times 5}{5} = \frac{805}{5}$

$n = 161$ miles

6.2 Exercises

1. hecto-

3. deci-

5. kilo-

7. 46 centimeters = 460 millimeters

9. 2.61 kilometers = 2610 meters

11. 1670 millimeters = 1.67 meters

13. 7.32 centimeters = 0.0732 meters

15. 2 kilometers = 200,000 centimeters

17. 78,000 millimeters = 0.078 kilometer

19. 35 mm = 3.5 cm = 0.035 m

21. 4.5 km = 4500 m = 450,000 cm

23. (b)

25. (c)

27. (b)

29. (a)

31. (b)

33. 390 decimeters = 39 meters

35. 800 dekameters = 8000 meters

37. 48.2 meters = 0.482 hectometer

39. 243 m + 2.7 km + 312 m

= 243 m +2700 m + 312 m

= 3255 m

41. 5.2 cm + 361 cm +968 mm

= 5.2 cm + 361 cm + 96.8 cm

= 463 cm

43. 15 mm + 2 dm +42 cm

= 1.5cm + 20 cm + 42 cm

= 63.5 cm

45. 0.95 cm + 1.35 cm + 2.464 mm

= 0.95 cm + 1.35 cm + 0.2464 cm

= 2.5464 cm or 25.464 mm

47. 65 cm + 80 mm + 2.5 m

= 0.65 m + 0.08 m + 2.5 m

= 3.23 m

49. $46 \text{ m} + 986 \text{ cm} + 0.884 \text{ km}$
$= 46 \text{ m} + 9.86 \text{ m} + 884 \text{ m}$
$= 939.86 \text{ m}$

51. 96.4 centimeters = 0.964 meter

53. False

55. True

57. True

59. True

61. a. 4818 meters = 481,800 centimeters
b. 4818 meters = 4.818 kilometers

63. 0.000000254 centimeters
= 0.00000000254 meters

65.
$$\begin{array}{r} 39.610,000 \\ -\ 31,940,000 \\ \hline 7,670,000 \end{array} \text{ metric tons}$$

67. $18,560,000 \times 1000 = 18,560,000,000$ kilograms

69.
$$\begin{array}{r} 36,980,000 \\ -\ \ 12,100,000 \\ \hline 24,880,000 \end{array}$$
$$\text{percent} = \frac{24,880,000}{12,100,000} \approx 2.0562 = 205.62\%$$
For 2010:
$3.0562 \times 36,980,000 \approx 113,020,000$ metric tons

Cumulative Review

71. $n = 0.03\% \times 5900$
$n = 0.0003 \times 5900$
$n = 1.77$

73. $n = 75\% \times 20$
$n = 0.75 \times 20$
$n = 15$

6.3 Exercises

1. 1 kL

3. 1 mg

5. 1g

7. 9 kL = 9000 L

9. 12L = 12,000 mL

11. 18.9 mL = 0.0189 L

13. 752 L = 0.752 kL

15. 2.43 kL = 2,430,000 mL

17. 82 mL = 82 cm^3

19. 5261 mL = 0.005261 kL

21. 74 L = 74,000mL = 74, 000 cm^3

23. 216 g = 0.216 kg

25. 35 mg = 0.035 g

27. 6328 mg = 6.328 g

29. 2.92 kg = 2920 g

31. 2.4 t = 2400 kg

33. 7 mL = 0.007 L = 0.00007 kL

35. 128 cm^3 = 0.128 L = 0.000128 kL

37. 0.033 kg = 33 g = 33,000 mg

39. 2.58 metric tons = 2580 kg = 2,580,000 g

41. (b)

43. (a)

45. 83 L + 822 mL + 30.1 L
= 83 L + 0.822 L + 30.1 L
= 113.922 L

47. 20 g + 52 mg + 1.5 kg
= 20 g + 0.052 g + 1500 g
= 1520.052 g

49. True

51. False

53. False

55. True

57. $\frac{n}{3.624 \text{ kg}} = \frac{\$8.99}{0.453 \text{ kg}}$

$n = \frac{3.624 \text{ kg} \times \$8.99}{0.453 \text{ kg}}$

$n = \$71.92$

59. $0.4 \text{ L} = 400 \text{ mL}$

$\frac{\$850}{\text{mL}} \times 400 \text{ mL} = \$340{,}000$

61. $0.45 \text{ t} = 450 \text{ kg}$

$450 \text{ kg} \times \frac{\$22{,}450}{\text{kg}} = \$10{,}102{,}500$

63. 5632 picograms = 0.005632 micrograms

65. $4100 - 1810 = 2290$ kg more

67. $\text{Value} = \frac{24{,}000{,}000}{2610}$

$\approx \$9195.40$ per kilogram
or \$9.20 per gram

69. 1997

Cumulative Review

71. $\text{percent} = \frac{14}{70}$

$= 0.20$

$= 20\%$

73. $n = 10\%$ of 48000

$n = 0.1 \times 4800$

$n = 480$

Price = 4800 − 480 = 4320

$n = 105\% \times 4320$

$n = 1.05 \times 4320$

$n = 4536$

\$4536

How Am I Doing? Sections 6.1-6.3

1. $52 \text{ inches} \times \frac{1 \text{ foot}}{12 \text{ inches}} = 4.33 \text{ feet}$

2. $24 \text{ quarts} \times \frac{1 \text{ gallon}}{4 \text{ quarts}} = 6 \text{ gallons}$

3. $3 \text{ miles} \times \frac{1720 \text{ yards}}{\text{miles}} = 5280 \text{ yards}$

4. $6400 \text{ pounds} \times \frac{1 \text{ ton}}{2000 \text{ pounds}} = 3.2 \text{ tons}$

5. $22 \text{ minutes} \times \frac{60 \text{ seconds}}{1 \text{ minute}} = 1320 \text{ minutes}$

6. $5 \text{ gallons} \times \frac{8 \text{ pints}}{1 \text{ gallon}} = 40 \text{ pints}$

7. $24 \text{ ounces} \times \frac{1 \text{ pound}}{16 \text{ ounces}} = 1.5 \text{ pounds}$

$1.5 \text{ pounds} \times \frac{\$5.50}{1 \text{ pound}} = \$8.25$

8. $6.75 \text{ km} = 6750 \text{ m}$

9. $73.9 \text{ m} = 7390 \text{ cm}$

10. $986 \text{ mm} = 98.6 \text{ cm}$

11. $27 \text{ mm} = 0.027\text{m}$

12. $5296 \text{ mm} = 529.6 \text{ cm}$

13. $482 \text{ m} = 0.482 \text{ km}$

14. $1.2 \text{ km} = 1200 \text{ m}$

$$\begin{array}{r} 1200 \\ 192 \\ +\ 984 \\ \hline 2376 \text{ m} \end{array}$$

15. $9342 \text{ mm} = 9.342 \text{ m}$

$3862 \text{ cm} = 38.62 \text{ m}$

$$\begin{array}{r} 9.342 \\ 38.620 \\ +\ 46.300 \\ \hline 94.262 \text{ m} \end{array}$$

16. $$\begin{array}{r} 78 \text{ cm} = 0.78 \text{ m} \\ +\ 128 \text{ cm} = 1.28 \text{ m} \\ \hline 2.06 \text{ m} \end{array}$$

$$\begin{array}{r} 3.40 \\ -\ 2.06 \\ \hline 1.34 \text{ m or } 134 \text{ cm} \end{array}$$

17. $5.66 \text{ L} = 5660 \text{ mL}$

18. $7835 \text{ g} = 7.835 \text{ kg}$

19. $56.3 \text{ kg} = 0.0563 \text{ t}$

20. $4.8 \text{ kL} = 4800 \text{ L}$

21. $568 \text{ mg} = 0.568 \text{ g}$

22. $8.9 \text{ L} = 8900 \text{ cm}^3$

23. $75 \text{ kg} = 75{,}000 \text{ g}$

$$\frac{x}{75{,}000} = \frac{7.75}{5000}$$
$$5000x = 7.75(75{,}000)$$
$$\frac{5000x}{5000} = \frac{581{,}250}{5000}$$
$$x = 116.25$$

\$116.25

24. $35 \text{ kg} = 35{,}000 \text{ g}$

$$\frac{x}{35{,}000} = \frac{32.50}{5000}$$
$$5000x = 32.50(35{,}000)$$
$$\frac{5000x}{5000} = \frac{1{,}137{,}500}{5000}$$
$$x = 227.5$$

\$227.50

25. $200 \text{ mL} = 0.2 \text{ L}$

$0.2 \text{ L} \times \frac{36}{1 \text{ L}} = \$7.20$

26. $0.5 \text{ kg} = 500 \text{ g}$

$500 \text{ g} \times \frac{\$11.20}{1 \text{ g}} = \$5600$

6.4 Exercises

1. The meter is approximately the same length as a yard. The meter is slightly longer.

3. The inch is approximatly twice the length of a centimeter.

5. $7 \text{ ft} \times \frac{0.305 \text{ m}}{1 \text{ ft}} \approx 2.14 \text{ m}$

7. $9 \text{ in.} \times \frac{2.54 \text{ cm}}{1 \text{ in.}} \approx 22.86 \text{ cm}$

9. $14\text{ m} \times \dfrac{1.09\text{ yd}}{1\text{ m}} \approx 15.26\text{ yd}$

11. $30.8\text{ yd} \times \dfrac{0.914\text{ m}}{1\text{ yd}} \approx 28.15\text{ m}$

13. $82\text{ mi} \times \dfrac{1.61\text{ km}}{1\text{ mi}} \approx 132.02\text{ km}$

15. $16\text{ m} \times \dfrac{1.09\text{ yd}}{1\text{ m}} \approx 17.44\text{ yd}$

17. $17.5\text{ cm} \times \dfrac{0.394\text{ in.}}{1\text{ cm}} \approx 6.90\text{ in.}$

19. $200\text{ m} \times \dfrac{3.28\text{ ft}}{1\text{ m}} \approx 656\text{ ft}$

21. $5\text{ km} \times \dfrac{0.62\text{ mi}}{1\text{ km}} \approx 3.1\text{ mi}$

23. $50\text{ gal} \times \dfrac{3.79\text{ L}}{1\text{ gal}} \approx 189.5\text{ L}$

25. $23\text{ qt} \times \dfrac{0.946\text{ L}}{1\text{ qt}} \approx 21.76\text{ L}$

27. $19\text{ L} \times \dfrac{0.264\text{ gal}}{1\text{ L}} \approx 5.02\text{ gal}$

29. $4.5\text{ L} \times \dfrac{1.06\text{ qt}}{1\text{ L}} \approx 4.77\text{ qt}$

31. $82\text{ kg} \times \dfrac{2.2\text{ lb}}{1\text{ kg}} \approx 180.4\text{ lb}$

33. $130\text{ lb} \times \dfrac{0.454\text{ kg}}{1\text{ lb}} \approx 59.02\text{ kg}$

35. $26\text{ oz} \times \dfrac{28.35\text{ g}}{1\text{ oz}} \approx 737.1\text{ g}$

37. $152\text{ kg} \times \dfrac{2.2\text{ lb}}{\text{kg}} \approx 334.4\text{ lb}$

39. $126\text{ g} \times \dfrac{0.0353\text{ oz}}{1\text{ g}} \approx 4.45\text{ oz}$

41. $1260\text{ cm} \times \dfrac{0.394\text{ in.}}{1\text{ cm}} \times \dfrac{1\text{ ft}}{12\text{ in.}} \approx 41.37\text{ ft}$

43. $\dfrac{55\text{ km}}{\text{hr}} \times \dfrac{0.62\text{ mi}}{1\text{ km}} = 34.1\text{ mi/hr}$

45. $\dfrac{400\text{ ft}}{1\text{ sec}} \times \dfrac{3600\text{ sec}}{1\text{ hr}} \times \dfrac{1\text{ mi}}{5280\text{ ft}} \approx 273\text{ mi/hr}$

47. $13\text{ mm} \times \dfrac{1\text{ cm}}{10\text{ mm}} \times \dfrac{0.394\text{ in.}}{1\text{ cm}} \approx 0.51\text{ in.}$

49.
$$\begin{aligned} F &= 1.8 \times C + 32 \\ &= 1.8 \times 85 + 32 \\ &= 185 \end{aligned}$$
185° F

51.
$$\begin{aligned} F &= 1.8 \times C + 32 \\ &= 1.8 \times 12 + 32 \\ &= 53.6 \end{aligned}$$
53.6° F

53.
$$\begin{aligned} C &= \frac{5 \times F - 160}{9} \\ &= \frac{5 \times 140 - 160}{9} \\ &= \frac{540}{9} \\ &= 60 \end{aligned}$$
60° C

55.
$$\begin{aligned} C &= \frac{5 \times F - 160}{9} \\ &= \frac{5 \times 40 - 160}{9} \\ &= \frac{40}{9} \\ &\approx 4.44 \end{aligned}$$
4.44° C

57. $\frac{65 \text{ miles}}{\text{hr}} \times \frac{1.61 \text{ km}}{\text{mile}} \approx 104.65 \text{ km/hr}$

Yes, she is speeding.

59. $15 \text{ gal} \times \frac{3.79 \text{ L}}{1 \text{ gal}} \approx 56.85 \text{ L}$

Difference $= 56.85 - 38 = 18.85$ L

61. $635 \text{ kg} \times \frac{2.2 \text{ lb}}{1 \text{ kg}} \approx 1397 \text{ lb}$

63. $F = 1.8 \times C + 32$

4 A.M.: $F = 1.8 \times 19 + 32 = 66.2$

7 A.M.: $F = 1.8 \times 45 + 32 = 113$

It is 66.2° F at 4 A.M.

It may reach 113° F after 7 A.M.

65. $96{,}550 \text{ km} \times \frac{0.62 \text{ mi}}{1 \text{ km}} \approx 59{,}861 \text{ mi}$

67. $28 \times 2.54 \times 2.54 \approx 180.6448$ sq cm

69. American: $8 \times 4 \times 28 = \$896$

German: $8 \text{ yd} \times \frac{0.914 \text{ m}}{1 \text{ yd}} \approx 7.312 \text{ m}$

$4 \text{ yd} \times \frac{0.914 \text{ m}}{1 \text{ yd}} \approx 3.656 \text{ m}$

$7.312 \times 3.656 \times 30 \approx \$802$

German carpet is cheaper by $896 - 802 = \$94$.

Cumulative Review

71.
$$\begin{aligned} 3^4 \times 2 - 5 + 12 &= 81 \times 2 - 5 + 12 \\ &= 162 - 5 + 12 \\ &= 169 \end{aligned}$$

73.
$$\begin{aligned} \frac{1}{2} \cdot \frac{3}{4} - \frac{1}{5}\left(\frac{1}{2}\right)^2 &= \frac{1}{2} \cdot \frac{3}{4} - \frac{1}{5}\left(\frac{1}{4}\right) \\ &= \frac{3}{8} - \frac{1}{20} \\ &= \frac{15}{40} - \frac{2}{40} \\ &= \frac{13}{40} \end{aligned}$$

6.5 Exercises

1. $14\frac{1}{3} + 8\frac{2}{3} + 13 = 36$ in.

$36 \text{ in.} \times \frac{1 \text{ ft}}{12 \text{ in.}} = 3 \text{ ft}$

3. $86 \text{ yd} + 77 \text{ yd} = 163 \text{ yd}$

$522 \text{ ft} \times \frac{1 \text{ yd}}{3 \text{ ft}} = 174 \text{ yd}$

$174 - 163 = 11$ yd left over

5. Length $= 200 + 90 + 200 = 490 \text{ cm} = 4.9 \text{ m}$

Cost $= 4.9 \text{ m} \times \frac{\$6.00}{1 \text{ m}} = \$29.40$

7. $1.836 \text{ km} = 1863 \text{ m}$

$\frac{1863 \text{ m}}{230} = 8.1$ m/space

9. $880 \text{ yd} \times \frac{0.914 \text{ m}}{1 \text{ yd}} = 803.32 \text{ m}$

880 yd is 4.32 m longer than 800 m

11. $\frac{\$0.46}{\text{L}} \times \frac{3.79 \text{ L}}{1 \text{ gal}} = \$1.74/\text{gal}$

Gasoline is more expensive in Mexico.

13.
$$\begin{aligned} F &= 1.8 \times C + 32 \\ &= 1.8 \times 25 + 32 \\ &= 45 + 32 \\ &= 77^\circ \text{ F} \end{aligned}$$

$86 - 77 = 9$

The difference is 9° F. The temperature in Boston was 9°F greater.

15.
$$\begin{aligned} F &= 1.8 \times 180 + 32 \\ &= 324 + 32 \\ &= 356^\circ \text{ F} \end{aligned}$$

$356 - 350 = 6$

The difference is 6° F. 180°C is hotter.

17. a. $\frac{520 \text{ mi}}{8 \text{ hr}} = 65 \text{ mi/hr}$

$\frac{65 \text{ mi}}{1 \text{ hr}} \times \frac{1.61 \text{ km}}{1 \text{ mi}} \approx 105 \text{ km/hr}$

b. Probably not. We cannot be sure, since they could speed for a short time, but we have no evidence to indicate that they broke the speed limit.

19. $2 \times \frac{1 \text{ pt}}{1 \text{ min}} \times \frac{1 \text{ qt}}{2 \text{ pt}} \times \frac{1 \text{ gal}}{1 \text{ qt}} \times \frac{60 \text{ min}}{1 \text{ hr}} = 15 \text{ gal/hour}$

21. $1.8 \text{ tons} \times \frac{2000 \text{ lb}}{1 \text{ ton}} \times \frac{\$0.03}{1 \text{ lb}} = \$108$

23. $14 \text{ oz} - 11 \text{oz} = 3 \text{ oz}$

$3 \text{ oz} \times \frac{28.35 \text{ g}}{1 \text{ oz}} \approx 85.05 \text{ g}$

25. a. $11 \text{ L} \times \frac{1.06 \text{ qt}}{1 \text{ L}} \approx 11.66 \text{ qt}$

$16 - 11.66 = 4.34 \text{ qt}$

He has 4.34 qt. extra.

b. $11.66 \text{ qt} \times \frac{\$2.89}{1 \text{ qt}} \approx \$33.70$

27. a. $392 \text{ km} \times \frac{1 \text{ L}}{6 \text{ km}} \times \frac{\$0.78}{1 \text{ L}} = \$5.46$

b. $\frac{56 \text{ km}}{1 \text{ L}} \times \frac{0.62 \text{ mi}}{1 \text{ km}} \times \frac{1 \text{ L}}{0.264 \text{ gal}} \approx 132 \text{ mi/gal}$

29. $\frac{240,000 \text{ gal}}{1 \text{ hr}} \times \frac{8 \text{ pt}}{1 \text{ gal}} \times \frac{1 \text{ hr}}{3,600 \text{ sec}} = 533\frac{1}{3} \text{ pt/sec}$

Yes

Cumulative Review

31. $\frac{18}{27} = \frac{n}{15}$

$18 \times 15 = 27 \times n$

$\frac{270}{27} = \frac{27 \times n}{27}$

$10 = n$

33. $6 \text{ in.} \times \frac{7.75 \text{ mi}}{3 \text{ in.}} = 15.5 \text{ mi}$

Putting Your Skills To Work

1. $(1 \text{ foot})^2 = (0.305 \text{ m})^2 \approx 0.093025 \text{ m}^2$

2. $43,560 \text{ ft}^2 \times \frac{0.093025 \text{ m}^2}{1 \text{ ft}^2}$

$= 4052.169 \text{ m}^2$ in 1 acre

3. $4 \text{ acres} \times \frac{0.405 \text{ hectare}}{1 \text{ acre}} = 1.62 \text{ hectares}$

4. $3 \text{ hectores} \times \frac{1 \text{ acre}}{0.405 \text{ hectores}} \approx 7.41 \text{ acres}$

5. $A = 340 \times 590 = 200,600 \text{ ft}^2$

$200,600 \text{ ft}^2 \times \frac{1 \text{ acre}}{43,560 \text{ ft}^2} \times \frac{0.405 \text{ hectare}}{1 \text{ acre}}$

≈ 1.87 hectares

6. $A = 200 \times 340 = 68,000 \text{ yd}^2$

$68,000 \text{ yd}^2 \times \frac{9 \text{ ft}^2}{1 \text{ yd}^2} = 612,000 \text{ ft}^2$

$612,000 \text{ ft}^2 \times \frac{1 \text{ acre}}{43,560 \text{ ft}^2} \times \frac{0.405 \text{ hectares}}{1 \text{ acre}}$

≈ 5.69 hectares

7. $2 \text{ hectores} \times \frac{1 \text{ acre}}{0.405 \text{ hectores}} \approx 4.94 \text{ acres}$

The error is $5.00 - 4.94 = 0.06$ acre.

8. $2 \text{ acres} \times \frac{0.405 \text{ hectares}}{1 \text{ acre}} = 0.81 \text{ hectare}$

It is $0.96 - 0.81 = 0.05$ hectares in error.

Chapter 6 Review Problems

1. $33 \text{ ft} \times \frac{1 \text{ yd}}{3 \text{ ft}} = 11 \text{ yd}$

2. $27 \text{ ft} \times \frac{1 \text{ yd}}{3 \text{ ft}} = 9 \text{ yd}$

3. $5 \text{ mi} \times \frac{1760 \text{ yd}}{1 \text{ mi}} = 8800 \text{ yd}$

4. $6 \text{ mi} \times \frac{1760 \text{ yd}}{1 \text{ mi}} = 8 = 10,560 \text{ yd}$

5. $90 \text{ in} \times \frac{1 \text{ ft}}{12 \text{ in}} = 7.5 \text{ ft}$

6. $78 \text{ in.} \times \frac{1 \text{ ft}}{12 \text{ in.}} = 6.5 \text{ ft}$

7. $15,840 \text{ ft} \times \frac{1 \text{ mi}}{5280 \text{ ft}} = 3 \text{ mi}$

8. $10,560 \text{ ft} \times \frac{1 \text{ mi}}{5280 \text{ ft}} = 2 \text{ mi}$

9. $7 \text{ tons} \times \frac{2000 \text{ lb}}{1 \text{ ton}} = 14,000 \text{ lb}$

10. $4 \text{ tons} \times \frac{2000 \text{ lb}}{1 \text{ ton}} = 8000 \text{ lb}$

11. $8 \text{ oz} \times \frac{1 \text{ lb}}{16 \text{ oz}} = 0.5 \text{ lb}$

12. $12 \text{ oz} \times \frac{1 \text{ lb}}{16 \text{ oz}} = 0.75 \text{ lb}$

13. $15 \text{ gal} \times \frac{4 \text{ qt}}{1 \text{ gal}} = 60 \text{ qt}$

14. $21 \text{ gal} \times \frac{4 \text{ qt}}{1 \text{ gal}} = 84 \text{ qt}$

15. $31 \text{ pt} \times \frac{1 \text{ qt}}{2 \text{ pt}} = 15.5 \text{ qt}$

16. $27 \text{ pt} \times \frac{1 \text{ qt}}{2 \text{ pt}} = 13.5 \text{ qt}$

17. 56 cm = 560 mm

18. 29 cm = 290 mm

19. 1763 mm = 176.3 cm

20. 2598 mm = 259.8 cm

21. 9.2 m = 920 cm

22. 7.4 m = 740 cm

23. 10,000 m = 10 km

24. 8200 m = 8.2 km

25. 6.2 m + 121 cm + 0.52 m
= 6.2m + 1.21m + 0.52m
= 7.93 m

26. 9.8 m + 673 cm + 0.48 m
= 9.8 m + 6.73 m + 0.48 m
= 17.01 m

27. 0.024 km + 1.8 m + 983 cm
= 24 m + 1.8 m + 9.82 m
= 35.63 m

28. 0.078 km + 5.5 m + 609 cm
= 78 m + 5.5 m + 6.09 m
= 89.59 m

29. 17 kL = 17,000 L

30. 23 kL = 23,000 L

31. 196 kg = 196,000 g

32. 721 kg = 721,000 g

33. 778 mg = 0.778 g

34. 459 mg = 0.459 g

35. 3500 g = 3.5 kg

36. 12,750 g = 12.75 kg

37. 765 cm^3 = 765 mL

38. 423 cm^3 = 423 mL

39. 2.43 L = 2430 mL = 2430 cm^3

40. 1.93 L = 1930 mL = 1930 cm^3

41. $42 \text{ kg} \times \frac{2.2 \text{ lb}}{1 \text{ kg}} \approx 92.4 \text{ lb}$

42. $9 \text{ ft} \times \frac{0.305 \text{ m}}{1 \text{ ft}} \approx 2.75 \text{ m}$

43. $45 \text{ mi} \times \frac{1.61 \text{ km}}{1 \text{ mi}} = 72.45 \text{ km}$

44. $88 \text{ mi} \times \frac{1.61 \text{ km}}{1 \text{ mi}} = 141.68 \text{ km}$

45. $14 \text{ cm} \times \frac{0.394 \text{ in}}{1 \text{ cm}} \approx 5.52 \text{ in.}$

46. $18 \text{ cm} \times \frac{0.394 \text{ in.}}{1 \text{ cm}} \approx 7.09 \text{ in.}$

47. $20 \text{ lb} \times \frac{0.454 \text{ kg}}{1 \text{ lb}} \approx 9.08 \text{ kg}$

48. $30 \text{ lb} \times \frac{0.454 \text{ kg}}{1 \text{ lb}} \approx 13.62 \text{ kg}$

49. $12 \text{ yd} \times \frac{0.914 \text{ m}}{1 \text{ yd}} \approx 10.97 \text{ m}$

50. $14 \text{ yd} \times \frac{0.914 \text{ m}}{1 \text{ yd}} \approx 12.80 \text{ m}$

51. $\frac{80 \text{ km}}{1 \text{ hr}} \times \frac{0.62 \text{ mi}}{1 \text{ km}} \approx 49.6 \text{ mi/hr}$

52. $\frac{70 \text{ km}}{1 \text{ hr}} \times \frac{0.62 \text{ mi}}{1 \text{ km}} \approx 43.4 \text{ mi/hr}$

53.
$$\begin{aligned} F &= 1.8 \times C + 32 \\ &= 1.8 \times 12 + 32 \\ &= 21.64 + 32 \\ &= 53.6^\circ \text{ F} \end{aligned}$$

54.
$$\begin{aligned} F &= 1.8 \times 32 + 32 \\ &= 57.6 + 32 \\ &= 89.6^\circ \text{ F} \end{aligned}$$

55.
$$\begin{aligned} C &= \frac{5 \times F - 160}{9} \\ &= \frac{5 \times 221 - 160}{9} \\ &= 105 \end{aligned}$$
105° C

56.
$$\begin{aligned} C &= \frac{5 \times F - 160}{9} \\ &= \frac{5 \times 185 - 160}{9} \\ &= 85 \end{aligned}$$
85° C

57.
$$\begin{aligned} C &= \frac{5 \times F - 160}{9} \\ &= \frac{5 \times 32 - 160}{9} \\ &= 0 \end{aligned}$$
0° C

58. $C = \dfrac{5 \times F - 160}{9}$
$= \dfrac{5 \times 212 - 160}{9}$
$= \dfrac{1060 - 160}{9}$
$= \dfrac{900}{9}$
$= 100^\circ\text{C}$

59. $13\text{ L} \times \dfrac{0.264\text{ gal}}{1\text{ L}} = 3.43\text{ gal}$

60. $27\text{ qt} \times \dfrac{0.946\text{ L}}{1\text{ qt}} = 25.54\text{ L}$

61. $5\text{ yd} \times \dfrac{3\text{ ft}}{\text{yd}} = 15\text{ ft}$
$A = lw$
$A = (18)(15)$
$A = 270\text{ sq ft}$
$270\text{ sq ft} \times \dfrac{1\text{ sq yd}}{9\text{ sq ft}} = 30\text{ sq yd}$

62. a. $7\frac{2}{3}\text{ ft} + 4\frac{1}{3}\text{ ft} + 5\text{ ft} = 16\frac{3}{3}\text{ ft}$
$=17\text{ feet}$

b. $17\text{ feet} \times \dfrac{12\text{ in.}}{1\text{ foot}} = 204\text{ in.}$

63. a. 16 m + 84 m + 16 m + 84 m = 200 m
b. 200 m = 0.2 km

64. $450\text{ g} \times \dfrac{0.0353\text{ oz}}{1\text{ g}} \approx 15.89\text{ oz}$
$\dfrac{\$0.14}{1\text{ oz}} \times 15.89\text{ oz} \approx \$2.22$

65. $15\text{ meters} \times \dfrac{3.28\text{ feet}}{1\text{ meter}} = 49.2\text{ feet}$
No; 4.2 feet short

66. $70\dfrac{\text{mi}}{\text{hr}} \times \dfrac{1.61\text{ km}}{1\text{ mi}} = 112.7\dfrac{\text{km}}{\text{hr}}$
Yes; she was driving 112.7 km/hr

67. $F = 1.8 \times C + 32$
$= 1.8 \times 185 + 32$
$= 365^\circ$
$= 390^\circ - 365^\circ = 25^\circ\text{F}$ too hot

68. $19\text{m} = 1900\text{ cm}$
Bottom part $= 1900 \times \dfrac{1}{5}$
$= 380\text{ cm}$

69. $\dfrac{100\text{ m}}{13.0\text{ sec}} \times \dfrac{3.28}{1\text{ m}} \approx 25.2\text{ ft/sec}$

70. $\dfrac{80\text{ km}}{1\text{ hr}} \times \dfrac{0.62\text{ mi}}{1\text{ km}} \approx 49.6\text{ mi/hr}$

71. $2.2 + 1.4 + 3.8 = 7.4\text{ kg}$
$7.4\text{ kg} \times \dfrac{2.2\text{ lb}}{1\text{ kg}} \approx 16.28\text{ lb}$
They are carrying 6.28 pounds; they are slightly over the weight limit.

72. $\dfrac{\$0.87}{1\text{ L}} \times \dfrac{1\text{ L}}{0.264\text{ gal}} \approx \$3.30/\text{gal}$

73. $1.88\text{ m} \times \dfrac{3.28\text{ ft}}{1\text{ m}} \approx 6.1664\text{ ft}$
Now, $6\text{ ft }2\text{ in} = 6\text{ ft}\dfrac{2}{12}\text{ ft}$
$= 6\frac{1}{6}\text{ ft}$
$\approx 6.1667\text{ ft}$
The person would be comfortable in the car.

74. $4\text{ m} \times \dfrac{3.28\text{ ft}}{1\text{ m}} \approx 13.12\text{ ft}$
$12\text{ m} \times \dfrac{3.28\text{ ft}}{1\text{ m}} \approx 39.36\text{ ft}$
Area $= 13.12 \times 39.36 \approx 516.4\text{ sq ft}$

75. $4 \times \dfrac{\$1.23}{1\text{ kg}} \times \dfrac{1\text{ kg}}{2.2\text{ lb}} \approx \$2.24$
The cost is about \$2.24.

How Am I Doing? Chapter 6 Test

1. $1.6 \text{ tons} \times \frac{2000 \text{ lb}}{1 \text{ ton}} = 3200 \text{ lb}$

2. $19 \text{ ft} \times \frac{12 \text{ in}}{1 \text{ ft}} = 228 \text{ in}$

3. $21 \text{ gal} \times \frac{4 \text{ qt}}{1 \text{ gal}} = 84 \text{ qt}$

4. $26{,}960 \text{ ft} \times \frac{1 \text{ mi}}{5280 \text{ ft}} = 7 \text{ mi}$

5. $1800 \text{ sec} \times \frac{1 \text{ min}}{60 \text{ sec}} = 30 \text{ min}$

6. $3 \text{ cups} \times \frac{1 \text{ qt}}{4 \text{ cups}} = 0.75 \text{ qt}$

7. $8 \text{ oz} \times \frac{1 \text{ lb}}{16 \text{ oz}} = 0.5 \text{ lb}$

8. $5.5 \text{ yd} \times \frac{3 \text{ ft}}{1 \text{ yd}} = 16.5 \text{ ft}$

9. 9.2 km = 9200 m

10. 9.88 cm = 0.0988 m

11. 46 mm = 4.6 cm

12. 12.7 m = 1270 cm

13. 0.936 cm = 9.36 mm

14. 46 L = 0.046 kL

15. 28.9 mg = 0.0289 g

16. 983 g = 0.983 kg

17. 0.92 L = 920 mL

18. 9.42 g = 9420 mg

19. $42 \text{ mi} \times \frac{1.61 \text{ km}}{1 \text{ mi}} \approx 67.62 \text{ km}$

20. $1.78 \text{ yd} \times \frac{0.914 \text{ m}}{1 \text{ yd}} \approx 1.63 \text{ m}$

21. $9 \text{ cm} \times \frac{0.394 \text{ in.}}{1 \text{ cm}} \approx 3.55 \text{ in.}$

22. $30 \text{ km} \times \frac{0.62 \text{ mi}}{1 \text{ km}} = 18.6 \text{ mi}$

23. $7.3 \text{ kg} \times \frac{2.2 \text{ lb}}{1 \text{ kg}} \approx 16.06 \text{ lb}$

24. $3 \text{ oz} \times \frac{28.35 \text{ g}}{1 \text{ oz}} \approx 85.05 \text{ g}$

25. $15 \text{ gal} \times \frac{3.79 \text{ L}}{1 \text{ gal}} \approx 56.85 \text{ L}$

26. $3 \text{ L} \times \frac{1.06 \text{ qt}}{1 \text{ L}} = 3.18 \text{ qt}$

27. a. $3 \text{ m} + 7 \text{ m} + 3 \text{ m} + 7 \text{ m} = 20 \text{ m}$

b. $20 \text{ m} \times \frac{1.09 \text{ yd}}{1 \text{ m}} \approx 21.8 \text{ yd}$

28. a. $F = 1.8 \times C + 32$

$= 1.8 \times 35 + 32$

$= 95$

$95^\circ \text{F} - 80^\circ \text{F} = 15^\circ \text{F}$

b. Yes; $80^\circ \text{F} < 95^\circ \text{F}$

29. $\frac{5.5 \text{ qt}}{1 \text{ min}} \times \frac{1 \text{ gal}}{4 \text{ qt}} \times \frac{60 \text{ min}}{1 \text{ hr}} = 82.5 \text{ gal/hr}$

30. a. $100 \times 3 = 300 \text{ km}$

b. $300 \text{ km} \times \frac{0.62 \text{ mi}}{1 \text{ km}} \approx 186 \text{ mi}$

$200 - 186 = 14$ mi farther

31. $1\frac{6}{16}+2\frac{2}{16}+1\frac{12}{16}=4\frac{20}{16}$
$=5\frac{4}{16}$
$=5\frac{1}{4}$ lb

32. $F=1.8\times C+32$
$=1.8\times 40+32$
$=104$
104°F

Cumulative Test for Chapters 1 - 6

1. $9824 - 3796 = 6028$

$$\begin{array}{r} 9824 \\ -\,3796 \\ \hline 6028 \end{array}$$

2.

$$\begin{array}{r} 608 \\ \times\,305 \\ \hline 3040 \\ 18240 \\ \hline 185{,}440 \end{array}$$

3. $32\overline{)8645}$ = 270

$$\begin{array}{r} 64 \\ 224 \\ 224 \\ \hline 5 \end{array}$$

270 R 5

4. $\frac{1}{7}+\frac{3}{14}+\frac{2}{21}=\frac{6}{42}+\frac{9}{42}+\frac{4}{42}$
$=\frac{6+9+4}{42}$
$=\frac{19}{42}$

5.

$$\begin{array}{r} 3\frac{1}{8} \\ -\,1\frac{3}{4} \\ \hline \end{array} \qquad \begin{array}{r} 2\frac{9}{8} \\ -\,1\frac{6}{8} \\ \hline 1\frac{3}{8} \end{array}$$

6. $0.5\times(9-3)^2\div\frac{2}{3}$
$=0.5\times 6^2\div\frac{2}{3}$
$=0.5\times 36\div\frac{2}{3}$
$=18\times\frac{3}{2}$
$=27$

7. $\frac{21}{35}\overset{?}{=}\frac{12}{20}$
$21\times 20\overset{?}{=}35\times 12$
$420=420$ True

8. $\frac{0.4}{n}=\frac{2}{30}$
$0.4\times 30=n\times 2$
$\frac{12}{2}=\frac{n\times 2}{2}$
$n=6$

9. $\frac{6.5\text{ cm}}{68\text{ g}}=\frac{20\text{ cm}}{n\text{ g}}$
$6.5\times n=68\times 20$
$6.5\times n=1360$
$\frac{6.5\times n}{6.5}=\frac{1360}{6.5}$
$n\approx 209.23$ g

10. What percent of 66 is 165?
$n\times 66=165$
$\frac{n\times 66}{66}=\frac{165}{66}$
$n=2.5$
$=250\%$

11. $n=15\%\times 800$
$n=0.15\times 00$
$n=120$

12. 0.5% of what number is 100?

$0.5\% \times n = 100$

$\frac{0.005 \times n}{0.005} = \frac{100}{0.005}$

$n = 20,000$

13. $38 \text{ qt} \times \frac{1 \text{ gal}}{4 \text{ qt}} = 9.5 \text{ gal}$

14. $2.5 \text{ tons} \times \frac{2000 \text{ lb}}{1 \text{ ton}} = 5,000 \text{ lb}$

15. $7 \text{ pt} \times \frac{1 \text{ qt}}{2 \text{ pt}} = 3.5 \text{ qt}$

16. $25 \text{ ft} \times \frac{12 \text{ in.}}{1 \text{ ft}} = 300 \text{ in.}$

17. 3.7 km = 3700 m

18. 62.8 g = 0.0628 kg

19. 0.79 L = 790 mL

20. 5 cm = 0.05 m

21. $42 \text{ lb} \times \frac{16 \text{ oz}}{1 \text{ lb}} = 672 \text{ oz}$

22. $C = \frac{5 \times F - 160}{9}$

$= \frac{5 \times 50 - 160}{9}$

$= \frac{250 - 160}{9}$

$= \frac{90}{9}$

$= 10^\circ\text{C}$

23. $28 \text{ gal} \times \frac{3.79 \text{ L}}{1 \text{ gal}} \approx 106.12 \text{ L}$

24. $96 \text{ lb} \times \frac{0.454 \text{ kg}}{1 \text{ lb}} \approx 43.58 \text{ kg}$

25. $30 \text{ in} \times \frac{2.54 \text{ cm}}{1 \text{ in}} = 76.2 \text{ cm}$

26. $9 \text{ mi} \times \frac{1.61 \text{ km}}{1 \text{ m}} \approx 14.49 \text{ km}$

27. 6 yd + 4 yd + 3 yd = 13 yd

$13 \text{ yd} \times \frac{0.914 \text{ m}}{1 \text{ yd}} \approx 11.88 \text{ m}$

28. $F = 1.8 \times C + 32$

$= 11.8 \times 15 + 32$

$= 59$

$= 59^\circ \text{F}$

$= 59^\circ - 15^\circ = 44^\circ \text{F}$

29. $\frac{100 \text{ km}}{1 \text{ hr}} \times 1\frac{1}{2} \text{ hr} = 150 \text{ km}$

$150 \text{ km} \times \frac{0.62 \text{ mi}}{1 \text{ km}} \approx 93 \text{ mi}$

$100 - 93 = 7$

He needs to travel 7 miles farther.

30. $2.5 \times 80 = 200 \text{ ft}$

$200 \text{ ft} \times \frac{1 \text{ yd}}{3 \text{ ft}} = 66.6$

Chapter 7

7.1 Exercises

1. An acute angle is an angle whose measure is between 0° and 90°.

3. Complementary angles are two angles that have a sum of 90°.

5. When two lines intersect, the two angles that are opposite each other are called vertical angles.

7. A transversal is a line that intersects two or more other lines at different points.

9. $\angle ABD$, $\angle CBE$

11. $\angle ABD$ and $\angle CBE$
$\angle DBC$ and $\angle ABE$

13. There are no complementary angles.

15. $\angle LOJ = 180^\circ - 90^\circ = 90^\circ$

17. $\angle JON = 90^\circ - 65^\circ = 25^\circ$

19. $\angle JOM = 90^\circ + 20^\circ = 110^\circ$

21. $\angle NOK = 65^\circ + 90^\circ = 155^\circ$

23. $90^\circ - 31^\circ = 59^\circ$

25. $180^\circ - 127^\circ = 53^\circ$

27. $\angle a = 180^\circ - 146^\circ = 34^\circ$

29. $\angle a = 123^\circ - 88^\circ = 35^\circ$

31. $\angle a = 85^\circ - 18^\circ - 42^\circ = 25^\circ$

33. $\angle b = 102^\circ$

$$\begin{aligned}\angle a &= \angle c \\ &= 180^\circ - 102^\circ \\ &= 78^\circ\end{aligned}$$

35. $\angle b = 38^\circ$

$$\begin{aligned}\angle a &= \angle c \\ &= 180^\circ - 38^\circ \\ &= 142^\circ\end{aligned}$$

37.

$$\begin{aligned}\angle a &= \angle c \\ &= 48^\circ \\ \angle b &= 180^\circ - 48^\circ \\ &= 132^\circ\end{aligned}$$

39.

$$\begin{aligned}\angle c &= \angle d \\ &= \angle a \\ &= 123^\circ \\ \angle b &= \angle c \\ &= \angle f \\ &= \angle g \\ &= 180^\circ - 123^\circ \\ &= 57^\circ\end{aligned}$$

41.

$$\begin{aligned}\angle x &= 90^\circ - 84^\circ \\ &= 6^\circ\end{aligned}$$

43. New angle is
$56^\circ - 7^\circ = 49^\circ$ north of east

Cumulative Review

45. $3.2 + 4.3 + 5.8 = 13.3$ miles
Then, $23 - 13.3 = 9.7$ miles remain.

47. Value
$= 4(1000) + 3(500) + 1(300) - 5(400)$
$= 4000 + 1500 + 300 - 2000$
$= 5500 + 300 - 2000$
$= 5800 - 2000$
$= 3800$
\$3800 increase

49. Percent miles $= 100 + 5 = 105\%$
Miles $= 105\%$ of $24 = 1.05 \times 24 = 25.2$ miles

7.2 Exercises

1. a. perpendicular
b. equal

3. multiply

5. $P = 2l + 2w$
$= 2(5.5 \text{ mi}) + 2(2 \text{ mi})$
$= 11 \text{ mi} + 4 \text{ mi}$
$= 15 \text{ mi}$

7. $P = 2l + 2w$
$= 2(9.3 \text{ ft}) + 2(2.5 \text{ ft})$
$= 18.6 \text{ ft} + 5 \text{ ft}$
$= 23.6 \text{ ft}$

9. $P = 4s$
$= 4(4.3 \text{ in.})$
$= 17.2 \text{ in.}$

11. $P = 2l + 2w$
$= 2(0.84 \text{ mm}) + 2(0.12 \text{ mm})$
$= 1.68 \text{ mm} + 0.24 \text{ mm}$
$= 1.92 \text{ mm}$

13. $P = 2l + 2w$
$= 2(4.28 \text{ km}) + 2(4.28 \text{ km})$
$= 8.56 \text{ km} + 8.56 \text{ km}$
$= 17.12 \text{ km}$

15. $3.2 \text{ ft} \times \frac{12 \text{ in.}}{1 \text{ ft}} = 38.4 \text{ in.}$
$P = 2l + 2w$
$= 2(38.4 \text{ in.}) + 2(48 \text{ in.})$
$= 76.8 \text{ in.} + 96 \text{ in.}$
$= 172.8 \text{ in.}$

17. $P = 4s$
$= 4(0.068 \text{ mm})$
$= 0.272 \text{ mm}$

19. $P = 4s$
$= 4(7.96 \text{ cm})$
$= 31.84 \text{ cm}$

21.
$$\begin{array}{r} 10\tfrac{1}{2} \\ 7 \\ 6 \\ 5 \\ 4\tfrac{1}{2} \\ +\ 2 \\ \hline p = 35 \text{ cm} \end{array}$$

23.
$$\begin{array}{r} 9 \text{ cm} \\ 13 \text{ cm} \\ 11 \text{ cm} \\ 13 \text{ cm} \\ 16 \text{ cm} \\ 41 \text{ cm} \\ 36 \text{ cm} \\ +\ 41 \text{ cm} \\ \hline 64 \text{ m} \\ P = 180 \text{ cm} \end{array}$$

25. $A = s^2$

$= (2.5)^2$

$= 6.25 \text{ ft}^2$

27. $A = lw$

$= 8 \text{ mi} \times 1.5 \text{ mi}$

$= 12 \text{ mi}^2$

29. $39 \text{ yd} \times \dfrac{3 \text{ ft}}{1 \text{ yd}} = 117 \text{ ft}$

$A = lw$

$= (117 \text{ ft})(19 \text{ ft})$

$= 1053 \text{ ft}^2$

31. a. $A = 21 \text{ m} \times 12 \text{ m} + 6 \text{ m} \times 7 \text{ m}$

$= 252 \text{ m}^2 + 42 \text{ m}^2$

$= 294 \text{ m}^2$

b.
```
     21 m
     12 m
     11 m
      6 m
      7 m
      6 m
      3 m
 +   12 m
 --------
     78 m
```

33. $A = 220 \text{ ft} \times 50 \text{ ft}$

$= 11{,}000 \text{ ft}^2$

$\text{Cost} = 11{,}000 \text{ ft}^2 \times \dfrac{\$12.00}{\text{ft}^2}$

$= \$132{,}000$

35. $P = 2l + 2w$

$= 2 \times 12.5 \text{ ft} + 2 \times 9.5 \text{ ft}$

$= 25 \text{ ft} + 19 \text{ ft}$

$= 44 \text{ ft}$

$\text{Cost} = \dfrac{\$1.35}{\text{foot}} \times 44 \text{ ft} = \$59.40$

37. a. 1×7, 2×6, 3×5, 4×4

four possible shapes.

b. 7 ft^2, 12 ft^2, 15 ft^2, 16 ft^2

c. The square garden measuring 4 ft on a side.

39. $A = lw + lw$

$= (24 \text{ ft})(11 \text{ ft})(7 \text{ ft})(8 \text{ ft})$

$= 288 \text{ ft}^2 + 56 \text{ ft}^2$

$= 344 \text{ ft}^2$

Cost of carpet

$= 344 \text{ ft}^2 \times \dfrac{\$14.50}{\text{yd}^2} \times \dfrac{1 \text{ yd}^2}{9 \text{ ft}^2}$

$\approx \$554.22$

$P = 12 \text{ ft} + 17 \text{ ft} + 8 \text{ ft} + 7 \text{ ft} + 20 \text{ ft} + 24 \text{ ft}$

$= 88 \text{ ft}$

Cost of binding

$= 88 \text{ ft} \times \dfrac{\$1.50}{\text{yd}} \times \dfrac{1 \text{ yd}}{3 \text{ ft}}$

$= \$44$

Total cost

$= \$544.22 + \$44$

$= \$598.22$

41. $A = lw = 8 \text{ m} \times (8 \text{ m} - 3 \text{ m})$

$= 8 \text{ m} \times 5 \text{ m}$

$= 40 \text{ m}^2$

$50 \text{ cm} = 0.5 \text{ m}$

$40 \text{ cm} = 0.4 \text{ m}$

$A = l \times w = (8 - 0.5) \times (8 - 3 - 0.4)$

$= 7.5 \text{ m} \times 4.6 \text{ m}$

$= 34.5 \text{ m}^2$

$40 \text{ m}^2 - 34.5 \text{ m}^2 = 5.5 \text{ m}^2$ less

Cumulative Review

43.
```
    156.8
     27.2
 +   39.3
 --------
    223.3
```

45.
```
     1076
 ×   20.3
 --------
     3228
    21520
 --------
 21,842.8
```

47. $A = lw$
$= 16\frac{1}{2}$
$= \frac{33}{2} \times \frac{64}{3}$
$= 352 \text{ ft}^2$
$\text{Cost} = \frac{\$0.75}{\text{ft}^2} \times 352 \text{ ft}^2 = \$264$

7.3 Exercises

1. adding

3. perpendicular

5. $P = 2(2.8 \text{ m}) + 2(17.3 \text{ m})$
$= 5.6 \text{ m} + 34.6 \text{ m}$
$= 40.2 \text{ m}$

7. $P = 2(9.2 \text{ in.}) + 2(15.6 \text{ in.})$
$= 18.4 \text{ in.} + 31.2 \text{ in.}$
$= 49.6 \text{ in.}$

9. $A = bh$
$= (17.6 \text{ m})(20.15 \text{ m})$
$= 354.64 \text{ m}^2$

11. $A = bh$
$= (28 \text{ yd})(21.5 \text{ yd})$
$= 602 \text{ yd}^2$

13. $P = 4(12 \text{ m})$
$= 48 \text{ m}$
$A = bh$
$= (12 \text{ m})(6 \text{ m})$
$= 72 \text{ m}^2$

15. $P = 4(2.4 \text{ ft})$
$= 9.6 \text{ ft}$
$A = bh$
$= (2.4 \text{ ft})(1.5 \text{ ft})$
$= 3.6 \text{ ft}^2$

17. $P = 13 \text{ m} + 20 \text{ m} + 15 \text{ m} + 34 \text{ m}$
$= 82 \text{ m}$

19. $P = 55 \text{ ft} + 135 \text{ ft} + 80.5 \text{ ft} + 75.5 \text{ ft}$
$= 346 \text{ ft}$

21. $A = \frac{h(b+B)}{2}$
$= \frac{(12 \text{ yd})(9.6 \text{ yd} + 10.2 \text{ yd})}{2}$
$= \frac{(12 \text{ yd})(19.8 \text{ yd})}{2}$
$= 118.8 \text{ yd}^2$

23. $A = \frac{h(b+B)}{2}$
$= \frac{(265 \text{ m})(280 \text{ m} + 300 \text{ m})}{2}$
$= \frac{(265 \text{ m})(580 \text{ m})}{2}$
$= 76{,}850 \text{ m}^2$

25. a. $A = 28 \text{ m} \times 16 \text{ m} + \frac{1}{2} \times 9 \text{ m} \times (28 \text{ m} + 32 \text{ m})$
$= 718 \text{ m}^2$
b. rectangle
c. trapezoid

27. a. $A = (12 \text{ ft} \times 5 \text{ ft}) + \frac{1}{2} \times 18 \text{ ft} \times (12 \text{ ft} + 21 \text{ ft})$
$= 357 \text{ ft}^2$
b. parallelogram
c. trapezoid

29. Top area $= bh$

$= (46\text{ yd})(49\text{ yd})$

$= 2254\text{ yd}^2$

Bottom area $= (46\text{ yd})(31\text{ yd})$

$= 1426\text{ yd}^2$

Total area $= 2254\text{ yd}^2 + 1426\text{ yd}^2$

$= 3680\text{ yd}^2$

$$\text{Cost} = 3680\text{ yd}^2 \times \frac{\$22}{\text{yd}^2} = \$80,960$$

31. $\text{Top area} = \dfrac{h(b+B)}{2}$

$= \dfrac{(0.71\times b)(b+2.65b)}{2}$

$= \dfrac{(0.71b)(3.65b)}{2}$

$= 1.29575b^2$

Bottom area $= 1.29575b^2$

Middle area $= lw$

$= (2.65b)(b)$

$= 2.65b^2$

Total area $= 2(1.29575b^2) + 2.65b^2$

$= 5.2415b^2$

Cumulative Review

33. $10\text{ yd} \times \dfrac{3\text{ ft}}{1\text{ yd}} = 30\text{ ft}$

35. $18\text{ m} = 1800\text{ cm}$

7.4 Exercises

1. right

3. Add the measures of the two known angles and subtract that value from 180°.

5. You could concluded that the lengths of all three sides of the triangle are equal.

7. True

9. True

11. False

13. True

15. $180^\circ - (36^\circ + 74^\circ) = 180^\circ - 110^\circ$

$= 70^\circ$

17. $180^\circ - (44.6^\circ + 52.5^\circ) = 180^\circ - 97.1^\circ$

$= 82.9^\circ$

19. $P = 18\text{ m} + 45\text{ m} + 55\text{ m} = 118\text{ m}$

21. $P = 45.25\text{ in.} + 35.75\text{ in.} + 35.75\text{ in.}$

$= 116.75\text{ in.}$

23. $P = 3l = 3\left(3\frac{1}{3}\text{ mi}\right)$

$= 3\left(\frac{10}{3}\text{ mi}\right)$

$= 10\text{ mi}$

25. $A = \dfrac{bh}{2}$

$= \dfrac{(9\text{ in})(12.5\text{ in})}{2}$

$= 56.25\text{ in}^2$

27. $A = \dfrac{bh}{2}$

$= \dfrac{(17.5\text{ cm})(9.5\text{ cm})}{2}$

$= 83.125\text{ cm}^2$

29. $A = \frac{1}{2}bh = \frac{1}{2}\left(3\frac{1}{2}\text{ yd}\right)\left(4\frac{1}{3}\text{ yd}\right)$

$= \frac{1}{2}\left(\frac{7}{2}\right)\left(\frac{13}{3}\right)$

$= \frac{91}{12}\text{ yd}^2$

$= 7\frac{7}{12}\text{ yd}^2$

31. Large triangle area $= \frac{bh}{2}$

$= \frac{(16 \text{ cm})(18 \text{ cm})}{2}$

$= 144 \text{ cm}^2$

Small triangle area $= \frac{bh}{2}$

$= \frac{(5 \text{ cm})(7 \text{ cm})}{2}$

$= 17.5 \text{ cm}^2$

Net area $= 144 \text{ cm}^2 - 17.5 \text{ cm}^2$

$= 126.5 \text{ cm}^2$

33. Top area $= \frac{bh}{2}$

$= \frac{(16 \text{ yd})(4.5 \text{ yd})}{2} = 36 \text{ yd}^2$

Bottom area $= lw$

$= (16 \text{ yd})(9.5 \text{ yd}) = 152 \text{ yd}^2$

Total area $= 36 \text{ yd}^2 + 152 \text{ yd}^2 = 188 \text{ yd}^2$

35. Area of side $= lw$

$= (30 \text{ ft})(15 \text{ ft})$

$= 450 \text{ ft}^2$

Area of front (or back) $= lw + \frac{bh}{2}$

$= (20 \text{ ft})(15 \text{ ft}) + \frac{(20 \text{ ft})(12 \text{ ft})}{2}$

$= 300 \text{ ft}^2 + 120 \text{ ft}^2$

$= 420 \text{ ft}^2$

Area of 4 sides (total area) $= 2(450 \text{ ft}^2) + 2(420 \text{ ft}^2)$

$= 900 \text{ ft}^2 + 840 \text{ ft}^2$

$= 1740 \text{ ft}^2$

37. Draw a vertical line through the middle of the wind to produce two identical triangles. For each triangle, $b = 22$ yd and $h = \frac{26 \text{ yd}}{2} = 13$ yd

$A = \frac{bh}{2}$

$= \frac{(18 \text{ yd})(13 \text{ yd})}{2}$

$= 117 \text{ yd}^2$

Total area $= 2(117 \text{ yd}^2)$

$= 234 \text{ yd}^2$

Cost $= 234 \text{ yd}^2 \times \frac{\$90}{\text{yd}^2}$

$= \$21,060$

39. Largest triangle area $= \frac{bh}{2}$

$= \frac{20h}{2}$

$= 10h$ m

Smallest triangle area $= \frac{bh}{2}$

$= \frac{1.25h}{2}$

$= 0.625h$ m

Percent $= \frac{0.625h}{10h}$

$= 0.0625$

$= 6.25\%$

Cumulative Review

41. $\frac{5}{n} = \frac{7.5}{18}$

$5 \times 18 = 7.5 \times n$

$\frac{90}{7.5} = \frac{7.5n}{7.5}$

$12 = n$

43. $\frac{4}{15} = \frac{n}{2685}$

$4 \times 2685 = 15 \times n$

$\frac{10,736}{15} = \frac{15 \times n}{15}$

$716 = n$

716 waitstaff

45. $\frac{n}{135} = \frac{14}{70}$

$70n = 135 \times 14$

$\frac{70n}{70} = \frac{1890}{70}$

$n = 27$ students

7.5 Exercises

1. $\sqrt{25} = 5$ since $(5)(5) = 25$

3. whole

5. To approximate the square root of a number that is not a perfect square, use the square root table or a calculator.

7. $\sqrt{9} = 3$

9. $\sqrt{64} = 8$

11. $\sqrt{0} = 0$

13. $\sqrt{144} = 12$

15. $\sqrt{169} = 13$

17. $\sqrt{100} = 10$

19. $\sqrt{49} + \sqrt{9} = 7 + 3 = 10$

21. $\sqrt{100} + \sqrt{1} = 10 + 1 = 11$

23. $\sqrt{225} - \sqrt{144} = 15 - 12 = 3$

25. $\sqrt{169} - \sqrt{121} + \sqrt{36} = 13 - 11 + 6$

$= 2 + 6$

$= 8$

27. $\sqrt{4} \times \sqrt{121} = 2 \times 11 = 22$

29. a. Yes because $16 \times 16 = 256$

b. $\sqrt{256} = 16$

31. $\sqrt{18} \approx 4.243$

33. $\sqrt{76} \approx 8.718$

35. $\sqrt{200} \approx 14.142$

37. $\sqrt{34 \text{ m}^2} \approx 5.831$ m

39. $\sqrt{136 \text{ m}^2} \approx 11.662$ m

41. $\sqrt{36} + \sqrt{20} \approx 6 + 4.472 = 10.472$

43. $\sqrt{198} - \sqrt{49} \approx 14.071 - 7 = 7.071$

45. $\sqrt{10,964}$ ft ≈ 104.7 ft

47. $\sqrt{16,200}$ ft ≈ 127.3 ft

49. a. $\sqrt{4} = 2$

b. $\sqrt{0.04} = 0.2$

c. Each answer is obtained from the previous answer by dividing by 10.

d. No because 0.004 is not a perfect square.

51. $\sqrt{456} + \sqrt{322}$

$\approx 21.3542 + 17.9443$

$= 32.285$ which rounds to 32.299

Cumulative Review

53. $A = lw = (60 \text{ in.})(80 \text{ in.}) = 4800$ sq in.

55. $30 \text{ km} \times \frac{1 \text{ mi}}{1.61 \text{ km}} = 18.6$ mi

How Am I Doing? Sections 7.1 - 7.5

1. $90^\circ - 72^\circ = 18^\circ$

2. $180^\circ - 63^\circ = 117^\circ$

3. $\angle a = 180^\circ - 136^\circ = 44^\circ$
   $\angle b = 136^\circ$
   $\angle a = \angle c = 44^\circ$

4. $P = 2(6.5) + 2(2.5)$
   $= 13 + 5$
   $= 18$ m

5. $P = 4(3.5) = 14$ m

6. $A = (4.8)^2 = 23.04$ sq cm

7. $A = 2.7 \times 0.9 = 2.43$ sq cm

8. $P = 2(9.2) + 2(3.6)$
   $= 18.4 + 7.2$
   $= 25.6$ yd

9. $P = 17 + 15 + 25\frac{1}{2} + 21\frac{1}{2}$
   $= 79$ ft

10. $A = 27 \times 13 = 351$ sq in.

11. $A = \frac{1}{2}(22 + 16)$
    $= \frac{1}{2}(38)(9)$
    $= 171$ sq in.

12. $A = 7 \times 9 + \frac{1}{2}(10 + 7)(4)$
    $= 63 + \frac{1}{2}(17)(4)$
    $= 97$ sq m

13. $180^\circ - (39^\circ + 118^\circ) = 180^\circ - 157^\circ = 23^\circ$

14. $P = 4\frac{2}{3} + 7\frac{1}{3} + 3$
    $= 11\frac{3}{3} + 3$
    $= 15$ m

15. $A = \frac{1}{2}(16)(9) = 72$ sq m

16. a. $A = \frac{1}{2}(44 + 30)(16) = \frac{1}{2}(74)(16)$
    $= 592$ sq ft of paint
    b. $P = 20 + 30 + 20 + 44 = 114$ ft of trim

17. $\sqrt{64} = 8$

18. $\sqrt{4} + \sqrt{100} = 2 + 10 = 12$

19. $\sqrt{169} = 13$

20. $\sqrt{256} = 16$

21. $\sqrt{46} = 6.782$

7.6 Exercises

1. Square the length of each leg and add these two results. Then take the square root of the result.

3. $h = \sqrt{12^2 + 9^2}$
   $= \sqrt{144 + 81}$
   $= \sqrt{225}$
   $= 15$ yd

5. $\text{leg} = \sqrt{16^2 - 5^2}$
   $= \sqrt{256 - 25}$
   $= \sqrt{231}$
   ≈ 15.199 ft

7. $h = \sqrt{11^2 + 3^2}$
$= \sqrt{121 + 9}$
$= \sqrt{130}$
≈ 11.402 m

9. $h = \sqrt{10^2 + 10^2}$
$= \sqrt{100 + 100}$
$= \sqrt{200}$
≈ 14.142 m

11. $\text{leg} = \sqrt{10^2 - 5^2}$
$= \sqrt{100 - 25}$
$= \sqrt{75}$
≈ 8.660 ft

13. $\text{leg} = \sqrt{14^2 - 10^2}$
$= \sqrt{196 - 100}$
$= \sqrt{96}$
≈ 9.798 yd

15. $h = \sqrt{12^2 + 9^2}$
$= \sqrt{144 + 81}$
$= \sqrt{225}$
$= 15$ m

17. $\text{leg} = \sqrt{16^2 - 11^2}$
$= \sqrt{256 - 121}$
$= \sqrt{135}$
≈ 11.620 ft

19. $\text{hypotenuse} = \sqrt{12^2 + 5^2}$
$= \sqrt{144 + 25}$
$= \sqrt{169}$
$= 13$ ft

21. $\text{hypotenuse} = \sqrt{9^2 + 4^2}$
$= \sqrt{81 + 16}$
$= \sqrt{97}$
≈ 9.8 cm

23. $\text{leg} = \sqrt{32^2 - 30^2}$
$= \sqrt{1024 - 900}$
$= \sqrt{124}$
≈ 11.1 yd

25. Side opposite $30^\circ = \frac{1}{2}(8) = 4$ in.
The other leg $= \sqrt{8^2 - 4^2}$
$= \sqrt{64 - 16}$
$= \sqrt{48}$
≈ 6.9 in.

27. The other leg = 6 m
hypotenuse $= \sqrt{2} \times$ leg
$= \sqrt{2} \times 6$
$\approx 1.414 \times 6$
≈ 8.5 m

29. The other leg = 18 cm
hypotenuse $= \sqrt{2} \times$ leg
$= \sqrt{2} \times 18$
$\approx 1.414 \times 18$
≈ 25.5 cm

31. $\text{leg} = \sqrt{10^2 - 7^2}$
$= \sqrt{100 - 49}$
$= \sqrt{51}$
$= 7.1$ in.

33. $h = \sqrt{(0.25)^2 + (0.4)^2}$
$= \sqrt{0.0625 + 0.16}$
$= \sqrt{0.2225}$
≈ 0.47 mi

35. $h = \sqrt{7^2 + 11^2}$
$= \sqrt{49 + 121}$
$= \sqrt{170}$
≈ 13.038 yd

Cumulative Review

37. $A = \frac{bh}{2}$
$= \frac{(31 \text{ m})(22 \text{ m})}{2}$
$= 341 \text{ m}^2$

39. $A = s^2$
$= (21 \text{ in.})^2$
$= 441 \text{ in.}^2$

7.7 Exercises

1. circumference

3. radius

5. You need to multiply the radius by 2 and then use the formula $C = \pi d$.

7. $d = 2r$
$= 2(29 \text{ in.})$
$= 58 \text{ in.}$

9. $d = 2r = 2\left(8\frac{1}{2}\right) = 17 \text{ mm}$

11. $r = \frac{d}{2}$
$= \frac{45 \text{ yd}}{2}$
$= 22.5 \text{ yd}$

13. $r = \frac{d}{2}$
$= \frac{19.84 \text{ cm}}{2}$
$= 9.92 \text{ cm}$

15. $C = \pi d$
$= 3.14(32 \text{ cm})$
$= 100.48 \text{ cm}$

17. $C = \pi r$
$= 2(3.14)(18.5 \text{ in.})$
$= 116.18 \text{ in.}$

19. $C = \pi d$
$= (3.14)(32 \text{ in.})$
$= 100.48 \text{ in.}$

$\text{Distance} = (100.48 \text{ in.})(5 \text{ rev}) \times \frac{1 \text{ ft}}{12 \text{ in.}}$
$\approx 41.9 \text{ ft}$

21. $A = \pi r^2$
$= 3.14(5 \text{ yd})^2$
$= 3.14(25 \text{ yd})^2$
$= 78.5 \text{ yd}^2$

23. $A = \pi r^2$
$= 3.14(8.5)^2$
$= 3.14(72.25)^2$
$\approx 226.87 \text{ in.}^2$

25. $r = \frac{d}{2} = \frac{32 \text{ cm}}{2} = 16 \text{ cm}$
$A = \pi r^2$
$= 3.14(16 \text{ cm})^2$
$= 3.14(256 \text{ cm}^2)$
$= 803.84 \text{ cm}^2$

27. $A = \pi r^2$
$= 3.14(12 \text{ ft})^2$
$= 3.14(144 \text{ ft}^2)$
$= 452.16 \text{ ft}^2$

29. $r = \frac{90 \text{ mi}}{2} = 45 \text{ mi}$
$A = \pi r^2$
$= 3.14(45 \text{ mi})^2$
$= 3.14(2025 \text{ mi}^2)$
$= 6358.5 \text{ mi}^2$

31. $A = \pi r^2 - \pi r^2$
$= 3.14(14\text{ m})^2 - 3.14(12\text{m})^2$
$= 3.14(196\text{ m}^2) - 3.14(144\text{ m}^2)$
$= 615.44\text{ m}^2 - 452.16\text{ m}^2$
$= 163.28\text{ m}^2$

33. $A = lw - \pi r^2$
$= (12\text{ m})(12\text{ m}) - (3.14)(6\text{ m})^2$
$= 144\text{ m}^2 - (3.14)(36\text{ m}^2)$
$= 144\text{ m}^2 - 113.04\text{ m}^3$
$= 30.96\text{ m}^2$

35. $r = \frac{d}{2} = \frac{10\text{ m}}{2} = 5\text{ m}$
$A = \frac{1}{2}\pi r^2 + lw$
$= \frac{1}{2}(3.14)(5\text{ m})^2 + (15\text{ m})(10\text{ m})$
$= \frac{1}{2}(3.14)(25\text{ m}^2) + 150\text{ m}^2$
$= 39.25\text{ m}^2 + 150\text{ m}^2$
$= 189.25\text{ m}^2$

37. $r = \frac{d}{2} = \frac{40\text{ yd}}{2} = 20\text{ yd}$
$A = \pi r^2 + lw$
$= (3.14)(20\text{ yd})^2 + (120\text{ yd})(40\text{ yd})$
$= 1256\text{ yd}^2 + 4800\text{ yd}^2$
$= 6056\text{ yd}^2$
$\text{Cost} = 6056.\text{ yd}^2 \times \frac{\$0.20}{\text{yd}^2} = \$1211.20$

39. $C = \pi d$
$= 3.14(3\text{ ft})$
$= 9.42\text{ ft}$

41. $C = 2\pi r$
$= 2(3.14)(30\text{ in})$
$= 188.4\text{ in}$
$\text{Distance} = (188.4)(9\text{rev}) \times \frac{1\text{ ft}}{12\text{ in}}$
$= 141.3\text{ ft}$

43. $1\text{ mi} = 5280\text{ ft} \times \frac{12\text{ in}}{1\text{ ft}}$
$= 63{,}360\text{ in.}$
$C = 2\pi r$
$= 2(3.14)(16\text{ in.})$
$= 100.48\text{ in.}$
$\text{rev} = \frac{63{,}360\text{ in.}}{100.48\text{ in.}}$
≈ 630.57

45. a. $C = \pi d = 3.14(8) = 25.12\text{ ft}$

b. $r = \frac{d}{2} = \frac{8}{2} = 4\text{ ft}$
$A = \pi r^2$
$= (3.14)(4)^2$
$= 50.24\text{ ft}^2$

47. $A = \pi r^2$
$= 3.14(200)^2$
$= 3.14(40{,}000)$
$= 125{,}600\text{ mi}^2$

49. a. $\text{Cost} = \frac{6.00}{8} \approx 0.75 = \$0.75$
$r = \frac{d}{2} = \frac{15\text{ in}}{2} = 7.5\text{ in.}$
$A = \pi r^2$
$= 3.14(7.5\text{ in.})^2$
$= 176.625\text{ in.}^2$
$\text{One slice} = \frac{176.625\text{ in.}^2}{8} \approx 22.1\text{ in.}^2$

b. $\text{Cost} = \frac{4.00}{6} \approx .67 = \$0.67$
$A = \pi r^2$
$= 3.14(6\text{ in.})^2$
$= 113.04\text{ in.}^2$
$\text{One slice} = \frac{113.04\text{ in.}^2}{6} \approx 18.84\text{ in.}^2$

c. Cost per square inch for 15 in. pizza is:

$$\frac{\$0.75}{22.1} \approx \$0.034.$$

Cost per square inch for 12 in. pizza is:

$$\frac{\$0.67}{18.84} \approx \$0.035.$$

The 15-inch pizza is a better value.

Cumulative Review

51. $n = 16\% \times 87$
$n = 16\% \times 87$
$n = 0.16 \times 87$
$n = 13.92$

53. $\frac{12}{100} = \frac{720}{b}$
$12 \times b = 100 \times 720$
$\frac{12 \times b}{12} = \frac{72{,}000}{12}$
$b = 6000$

55. $100\% - 40\% = 60\%$
What is 60% of 80
$n = 60\% \times 80$
$n = 0.6 \times 80$
$n = 48$ students

7.8 Exercises

1. a. sphere

b. $V = \frac{4\pi r^3}{3}$

3. a. cylinder

b. $V = \pi r^2 h$

5. a. cone

b. $V = \frac{\pi r^2 h}{3}$

7. $V = lwh$
$= (30 \text{ mm})(12 \text{ mm})(1.5 \text{ mm})$
$= 540 \text{ mm}^3$

9. $V = \pi r^2 h$
$= 3.14(3 \text{ m})^2(8 \text{ m})$
$= 3.14(9 \text{ m}^2)(8 \text{ m})$
$\approx 226.1 \text{ m}^3$

11. $r = \frac{d}{2} = \frac{22 \text{ m}}{2} = 11 \text{ m}$
$V = \pi r^2 h$
$= 3.14(11 \text{ m})^2(17 \text{ m})$
$= 3.14(121 \text{ m}^2)(17 \text{ m})$
$\approx 6459.0 \text{ m}^3$

13. $V = \frac{4\pi r^3}{3}$
$= \frac{4(3.14)(9 \text{ yd})^3}{3}$
$= \frac{4(3.14)(729 \text{ yd}^3)}{3}$
$\approx 3052.1 \text{ yd}^3$

15. $V = \frac{Bh}{3} = \frac{30(70)}{3} = 700 \text{ ft}^3$

17. $V = s^3 = (0.6)^3 = 0.216 \text{ cm}^3$

19. $V = \frac{\pi r^2 h}{3}$
$= \frac{3.14(3)^2(7)}{3}$
$= \frac{3.14(9)(7)}{3}$
$= 65.94 \text{ yd}^3$

21. $V = \frac{1}{2} \times \frac{4\pi r^3}{3}$
$= \frac{1}{2} \times \frac{4(3.14)(7\text{ m})^3}{3}$
$= \frac{1}{2} \times \frac{4(3.14)(343\text{ m}^3)}{3}$
$\approx \frac{1}{2} \times 1436.027\text{ m}^3$
$\approx 718.0\text{ m}^3$

23. $V = \frac{\pi r^2 h}{3}$
$= \frac{3.14(8\text{ cm})^2(14\text{ cm})}{3}$
$= \frac{3.14(64\text{ cm}^2)(14\text{ cm})}{3}$
$\approx 937.8\text{ cm}^3$

25. $V = \frac{\pi r^2 h}{3}$
$= \frac{3.14(5\text{ ft})^2(10\text{ ft})}{3}$
$= \frac{3.14(25\text{ ft}^2)(10\text{ ft})}{3}$
$\approx 261.7\text{ ft}^3$

27. $B = (7\text{ m})(7\text{ m}) = 49\text{ m}^2$
$V = \frac{Bh}{3}$
$= \frac{(49\text{ m}^2)(10\text{ m})}{3}$
$= 163.3\text{ m}^3$

29. $B = (8\text{ m})(14\text{ m}) = 112\text{ m}^2$
$V = \frac{Bh}{3}$
$= \frac{(112\text{ m}^2)(10\text{ m})}{3}$
$\approx 373.3\text{ m}^3$

31. $4\text{ in.} \times \frac{1\text{ ft}}{12\text{ in.}} \times \frac{1\text{ yd}}{3\text{ ft}} = \frac{1}{9}\text{ yd}$

$V = lwh$
$= (20\text{ yd})(18\text{ yd})\left(\frac{1}{9}\text{ yd}\right)$
$= (360\text{ yd}^2)\left(\frac{1}{9}\text{ yd}\right)$
$= 40\text{ yd}^3$

33. Outer $= \pi r^2 h$
$= 3.14(5\text{ in.})^2(20\text{ in.})$
$= 3.14(25\text{ in.}^2)(20\text{ in.})$
$= 1570\text{ in.}^3$
Inner $= \pi r^2 h$
$= 3.14(3\text{ in.})^2(20\text{ in.})$
$= 3.14(9\text{ in.}^2)(20\text{ in.})$
$= 565\text{ in.}^3$
Difference $= 1570\text{ in.}^3 - 565.2\text{ in.}^3$
$= 1004.8\text{ in.}^3$

35. Jupiter:

$V = \frac{4\pi r^3}{3}$
$= \frac{4(3.14)(45{,}000\text{ mi})^3}{3}$
$= 381{,}510{,}000{,}000{,}000\text{ mi}^3$

Earth:

$V = \frac{4\pi r^3}{3}$
$= \frac{4(3.14)(3950\text{ mi})^3}{3}$
$\approx 258{,}023{,}743{,}333\text{ mi}^3$

Difference

$= 381{,}510{,}000{,}000{,}000\text{ mi}^3$
$= -\quad 258{,}023{,}743{,}333\text{ mi}^3$
$381{,}251{,}976{,}256{,}667\text{ mi}^3$

37. Smaller

$$\begin{aligned} V &= lwh \\ &= (18 \text{ in.})(6 \text{ in.})(12 \text{ in.}) \\ &= (108 \text{ in.})(12 \text{ in.}) \\ &= 1296 \text{ in.}^3 \end{aligned}$$

Larger:

$$\begin{aligned} V &= lwh \\ &= (12 \text{ in.})(22 \text{ in.})(16 \text{ in.}) \\ &= (264 \text{ in.})(16 \text{ in.}) \\ &= 4224 \text{ in.}^3 \end{aligned}$$

Difference

$$\begin{aligned} &= 4224 \text{ in.}^3 - 1296 \text{ in.}^3 \\ &= 2928 \text{ in.}^3 \end{aligned}$$

39. $$\begin{aligned} V &= \frac{\pi r^2 h}{3} \\ &= \frac{3.14(6 \text{ cm})^2(10 \text{ cm})}{3} \\ &= \frac{3.14(36 \text{ cm}^2)(10 \text{ cm})}{3} \\ &= 376.8 \text{ cm}^3 \end{aligned}$$

$$\begin{aligned} \text{Cost} &= 376.8 \text{ cm}^3 \times \frac{\$3.00}{1 \text{ cm}^3} \\ &= \$1130.40 \end{aligned}$$

41. $h = 13.5 - 1.4 = 12.1$ cm

$$r = \frac{d}{2} = \frac{6.6}{2} = 3.3 \text{ cm}$$

$$\begin{aligned} V &= \pi r^2 h \approx 3.14(3.3)^2 \times 12.1 \\ &\approx 3.14(10.89)(12.1) \\ &\approx 413.8 \text{ cm}^3 \end{aligned}$$

43. $$\begin{aligned} B &= (87 \text{ yd})(130 \text{ yd}) \\ &= 11,310 \text{ yd}^3 \\ V &= \frac{Bh}{3} \\ &= \frac{(11,310 \text{ yd}^2)(70 \text{ yd})}{3} \\ &= 263,900 \text{ yd}^3 \end{aligned}$$

45. $$\begin{aligned} V &= (\text{base})(\text{height}) \\ &= \left(\frac{bh}{2}\right)(H) \\ &= \left(\frac{bhH}{2}\right) \text{ or } \frac{1}{2}hbH \end{aligned}$$

Cumulative Review

47.

$$\begin{array}{r} 7\frac{1}{3} \\ +\ 2\frac{1}{4} \\ \hline \end{array} \qquad \begin{array}{r} 7\frac{4}{12} \\ +\ 2\frac{3}{12} \\ \hline 9\frac{7}{12} \end{array}$$

49. $$\begin{aligned} 2\frac{1}{4} \times 3\frac{3}{4} &= \frac{9}{4} \times \frac{15}{4} \\ &= \frac{135}{16} \\ &= 8\frac{7}{16} \end{aligned}$$

51. $$\begin{aligned} &\left(\frac{5}{8} - \frac{1}{4}\right)^2 + \frac{7}{32} \\ &= \left(\frac{5}{8} - \frac{2}{8}\right)^2 + \frac{7}{32} \\ &= \left(\frac{3}{8}\right)^2 + \frac{7}{32} \\ &= \frac{9}{64} + \frac{14}{32} \\ &= \frac{23}{64} \end{aligned}$$

7.9 Exercises

1. size; shape

3. sides

5. $\frac{n}{2}=\frac{12}{3}$
$3n=(2)(12)$
$3n=24$
$\frac{3n}{3}=\frac{24}{3}$
$n=8$

8 m

7. $\frac{n}{6}=\frac{9}{21}$
$21n=(6)(9)$
$21n=54$
$\frac{21n}{21}=\frac{54}{21}$
$n\approx 2.6$

2.6 ft

9. $\frac{n}{8.5}=\frac{2}{5}$
$5n=8.5(2)$
$\frac{5n}{5}=\frac{17}{5}$
$n=3.4$ yd

11. *a* corresponds to *f*
b corresponds to *e*
c corresponds to *d*

13. $\frac{n}{8}=\frac{10.5}{25}$
$25n=(8)(10.5)$
$25n=84$
$\frac{25n}{25}=\frac{84}{25}$
$n\approx 3.4$

3.4 m

15. $3\text{ in.}\times\frac{1\text{ ft}}{12\text{ in.}}=\frac{1}{4}\text{ ft}$
$5\text{ in.}\times\frac{1\text{ ft}}{12\text{ in.}}=\frac{5}{12}\text{ ft}$
$\frac{\text{n}}{\frac{1}{4}}=\frac{3.5}{\frac{5}{12}}$
$\frac{5}{12}n=\left(\frac{1}{4}\right)(3.5)$
$\frac{5}{12}n=\frac{7}{8}$
$\frac{5}{12},\frac{12}{5}n=\frac{7}{8},\frac{12}{5}$
$n\approx 2.1$

2.1 ft

17. $\frac{n}{2}=\frac{5.5}{5}$
$5n=2\times 5.5$
$\frac{5n}{5}=\frac{11}{5}$
$n=2.2$

2.2 ft

19. $\frac{n}{6}=\frac{24}{4}$
$4n=(6)(24)$
$4n=144$
$\frac{6n}{6}=\frac{144}{4}$
$n=36$

36 ft

21. $\frac{n}{96}=\frac{5.5}{6.5}$
$6.5n=(96)(5.5)$
$6.5n=528$
$\frac{6.5n}{6.5}=\frac{528}{6.5}$
$n\approx 81$

81 ft

23. $\frac{n}{15}=\frac{5}{9}$

$9n=(15)(5)$

$9n=75$

$\frac{9n}{9}=\frac{75}{9}$

$n\approx 8.3$ ft

25. $\frac{n}{9}=\frac{8}{6}$

$6n=9\times 8$

$\frac{6n}{6}=\frac{72}{6}$

$n=12$ cm

27. $\left(\frac{2}{3}\right)^2=\frac{4}{9}$

Proportion:

$\frac{A}{26}=\frac{4}{9}$

$9A=(26)(4)$

$9A=104$

$\frac{9A}{9}=\frac{104}{9}$

$A\approx 11.6$

11.6 yd^2

Cumulative Review

29. $2\times 3^2+4-2\times 5$

$=2\times 9+4-2\times 5$

$=18+4-10$

$=22-10$

$=12$

31. $(5)(9)-(21+3)\div 8$

$=(5)(9)-(24)\div 8$

$=45-3$

$=42$

7.10 Exercises

1. a. Trip = 13 km + 17 km

= 30 km

$\text{Speed}=\frac{30\text{ km}}{0.4\text{ hr}}$

= 75 km/hr

b. Trip = 12 km + 15 km + 11 km

= 38 km

$\text{Speed}=\frac{38\text{ km}}{0.5\text{ hr}}$

= 76 km/hr

c. through Woodville and Palermo

3. $A=(7\text{ ft})(10\text{ ft})=70\text{ ft}^2$

$A=(7\text{ ft})(14\text{ ft})=98\text{ ft}^2$

$A=(6\text{ ft})(10\text{ ft})=60\text{ ft}^2$

$A=(6\text{ ft})(8\text{ ft})=48\text{ ft}^2$

Total A

$=70\text{ ft}^2+98\text{ ft}^2+60\text{ ft}^2+48\text{ ft}^2$

$=276\text{ ft}^2$

Time

$=276\text{ ft}^2\times\frac{15\text{ min}}{120\text{ ft}^2}=34.5\text{ min}$

5. $A=2(55\times 24)=2(1320)=2640\text{ ft}^2$

$A=2(32\times 24)=2(768)=1536\text{ ft}^2$

$A=16(4\times 2)=16(8)=128\text{ ft}^2$

$A=2(7\times 3)=2(21)=42\text{ ft}^2$

$\text{Total}=2640+1536-128-42$

$=4006\text{ ft}^2$

7. $A=lw-\frac{bh}{2}$

$=(21\text{ ft})(15\text{ ft})-\frac{(3\text{ ft})(6\text{ ft})}{2}$

$=315\text{ ft}^2-9\text{ ft}^2$

$=306\text{ ft}^2$

$\text{Cost}=306\text{ ft}^2\times\frac{1\text{ yd}^2}{9\text{ ft}^2}\times\frac{\$15}{\text{yd}^2}$

$=\$510$

9. $V = \frac{1}{2} \times \frac{4}{3} \times 3.14 \times (1 \text{ mm})^3$
$+ 3.14 \times 2 \text{ mm} \times (1 \text{ mm})^2$
$V \approx 8.37 \text{ mm}^3$
$\text{Cost} = 8.37 \text{ mm}^3 \times \frac{\$95}{\text{mm}^3}$
$= \$795.15$

11. a. $C = 2\pi r$
$= 2(3.14)(6500 \text{ km})$
$= 40,820 \text{ km}$

b. $S = \frac{40,820 \text{ km}}{2 \text{ hr}}$
$= 20,410 \text{ km/hr}$

13. $V = 400 \times \pi r^2 h$
$= 400 \times (3.14)(2 \text{ in.})^2 (10 \text{ in.})$
$= 400 \times (3.14)(4 \text{ in.})^2 (10 \text{ in.})$
$= 400 \times 125.6 \text{ in.}^3$
$\approx 50,240 \text{ in.}^3$

15. $P = 2l + 2w$
$= 2(456 \text{ ft}) + 2(625 \text{ ft})$
$= 912 \text{ ft} + 1250 \text{ ft}$
$= 2162 \text{ ft}$
$\text{Cost} = 2162 \text{ ft} \times \frac{1 \text{ yd}}{3 \text{ ft}} \times \frac{\$4.57}{\text{yd}}$
$\approx \$3293.45$

17. $r = \frac{d}{2} = \frac{28 \text{ in.}}{2} = 14 \text{ in.}$

Convert to feet:

$14 \text{ in.} \times \frac{1 \text{ ft}}{12 \text{ in.}} = \frac{14}{12} \text{ ft} = \frac{7}{6} \text{ ft}$

$C = 2\pi r = 2(3.14)\left(\frac{7}{6} \text{ ft}\right)$
$= \frac{21.98}{3} \text{ ft}$

Speed
$= (\text{rev})(C)$
$= (200 \text{ rev/min})\left(\frac{21.98}{3} \text{ ft}\right)$
$\approx 1465.33 \text{ ft/min}$

1465.33 ft/min

Cumulative Review

19.
$$\begin{array}{r} 128 \\ 16\overline{)2048} \\ \underline{16} \\ 44 \\ \underline{32} \\ 128 \\ \underline{128} \\ 0 \end{array}$$

21.
$$\begin{array}{r} 0.25 \\ 1.3\overline{)0.325} \\ \underline{26} \\ 65 \\ \underline{65} \\ 0 \end{array}$$

Putting Your Skills to Work

1. $d = \frac{460,000}{9.76} \approx 47,131 \text{ people/sq mi}$

2. $d = \frac{5,368,600}{676,367} \approx 8 \text{ people/sq mi}$

3. Answers will vary.

4. Alaska: $\frac{643,786}{571,951} \approx \frac{1 \text{ person}}{\text{sq. mi}}$

California: $\frac{35,116,033}{155,959} \approx 225 \frac{\text{people}}{\text{sq. mi}}$

Delaware: $\frac{807,385}{1954} \approx 413 \frac{\text{people}}{\text{sq. mi}}$

New Mexico: $\frac{1,855,089}{121,356} \approx 15 \frac{\text{people}}{\text{sq. mi}}$

Rhode Island: $\frac{1,069,725}{1045} \approx 1024 \frac{\text{people}}{\text{sq. mi}}$

5. $(1 \text{ mi})^2 = \left(1 \text{ mi} \times \frac{1760 \text{ yd}}{1 \text{ mi}}\right)^2$
$= 3,097,600 \text{ yd}^2$

$(1 \text{ mi})^2 = \left(1 \text{ mi} \times \frac{5280 \text{ ft}}{1 \text{ mi}}\right)^2$
$= 27,878,400 \text{ ft}^2$

6. $\frac{1 \text{ mi}^2}{47,131 \text{ people}} \times \frac{3,097,600 \text{ yd}^2}{1 \text{ mi}^2} \approx 66 \text{ yd}^2/\text{person}$

$\frac{1 \text{ mi}^2}{47,131 \text{ people}} \times \frac{27,878,400 \text{ ft}^2}{1 \text{ mi}^2} \approx 592 \text{ ft}^2/\text{person}$

7. $\frac{1 \text{ mi}^2}{1024 \text{ people}} \times \frac{3,097,600 \text{ yd}^2}{1 \text{ mi}^2} \approx 3025 \text{ yd}^2/\text{person}$

$\frac{1 \text{ mi}^2}{1024 \text{ people}} \times \frac{27,878,400 \text{ ft}^2}{1 \text{ mi}^2} \approx 27,225 \text{ ft}^2/\text{person}$

Chapter 7 Review Problems

1. $90^\circ - 76^\circ = 14^\circ$

2. $180^\circ - 76^\circ = 104^\circ$

3. $\angle b = 146^\circ$
$\angle a = \angle c$
$= 180^\circ - 146^\circ$
$= 34^\circ$

4. $\angle t = \angle x = \angle y = 65^\circ$
$\angle s = \angle u = \angle w = \angle z$
$= 180^\circ - 65$
$= 115^\circ$

5. $P = 2(8.3 \text{ m}) + 2(1.6 \text{ m})$
$= 16.6 \text{ m} + 3.2 \text{ m}$
$= 19.8 \text{ m}$

6. $P = 4s$
$= 4(2.4 \text{ yd})$
$= 9.6 \text{ yd}$

7. $A = (5.9 \text{ cm})(2.8 \text{ cm})$
$= 16.52 \text{ cm}^2$
$\approx 16.5 \text{ cm}^2$

8. $A = s^2$
$= (7.2 \text{ in.})^2$
$= 51.84 \text{ in.}^2$
$\approx 51.8 \text{ in.}^2$

9. $P = 3(8 \text{ ft}) + 2(2 \text{ ft}) + 4 \text{ ft} + 2(3 \text{ ft})$
$= 24 \text{ ft} + 4 \text{ ft} + 4 \text{ ft} + 6 \text{ ft}$
$= 38 \text{ ft}$

10. $P = 3(11 \text{ ft}) + 2(7 \text{ ft}) + 2(3.5 \text{ ft}) + 4 \text{ ft}$
$= 33 \text{ ft} + 14 \text{ ft} + 7 \text{ ft} + 4 \text{ ft}$
$= 58 \text{ ft}$

11. $A = (14 \text{ m})(5 \text{ m}) - 2(1 \text{ m})^2$
$= 70 \text{ m}^2 - 2 \text{ m}^2$
$= 68 \text{ m}^2$

12. $A = (9 \text{ m})^2 - (2.7 \text{ m})(6.5 \text{ m})$
$= 81 \text{ m}^2 - 17.55 \text{ m}^2$
$= 63.45 \text{ m}^2$
$\approx 63.5 \text{ m}^2$

13. $P = 2(52\text{ m}) + 2(20.6\text{ m})$
$= 104\text{ m} + 41.2\text{ m}$
$= 145.2\text{ m}$

14. $P = 5\text{ mi} + 22\text{ mi} + 5\text{ mi} + 30\text{ mi}$
$= 62\text{ mi}$

15. $A = (90\text{ m})(30\text{ m})$
$= 2700\text{ m}^2$

16. $A = \dfrac{36\text{ yd}(17\text{ yd} + 23\text{ yd})}{2}$
$= \dfrac{36\text{ yd}(40\text{ yd})}{2}$
$= 720\text{ yd}^2$

17. $A = \dfrac{(8\text{ cm})(33\text{ cm} + 20\text{ cm})}{2} + \dfrac{(20\text{ cm})(9\text{ cm} + 20\text{ cm})}{2}$
$= \dfrac{(9\text{ cm})(33\text{ cm})}{2} + \dfrac{(20\text{ cm})(29\text{ cm})}{2}$
$= 132\text{ cm}^2 + 290\text{ cm}^2$
$= 422\text{ cm}^2$

18. $A = (15\text{ m})(17\text{ m}) + (17\text{ m})(6\text{ m})$
$= 255\text{ m}^2 + 102\text{ m}^2$
$= 357\text{ m}^2$

19. $P = 18 + 21 + 21 = 60\text{ ft}$

20. $P = 15.5 + 15.5 + 15.5 = 46.5\text{ ft}$

21. $180^\circ - (15^\circ + 12^\circ) = 180^\circ - 27^\circ$
$= 153^\circ$

22. $180^\circ - (90^\circ + 35^\circ) = 180^\circ - 125^\circ$
$= 55^\circ$

23. $A = \dfrac{(8.5\text{ m})(12.3\text{ m})}{2}$
$= \dfrac{104.55\text{ m}^2}{2}$
$= 52.275\text{ m}^2$
$\approx 52.3\text{ m}^2$

24. $A = \dfrac{(12.5\text{ m})(9.5\text{ m})}{2}$
$= \dfrac{118.75\text{ m}^2}{2}$
$= 59.375\text{ m}^2$
$\approx 59.4\text{ m}^2$

25. $A = (18\text{ m})(22\text{ m}) + \dfrac{(18\text{ m})(6\text{ m})}{2}$
$= 396\text{ m}^2 + 54\text{ m}^2$
$= 450\text{ m}^2$

26. $A = (12\text{ m})(6\text{ m}) + \dfrac{(6\text{ m})(3\text{ m})}{2} + \dfrac{(6\text{ m})(2\text{ m})}{2}$
$= 72\text{ m}^2 + 9\text{ m}^2 + 6\text{ m}^2$
$= 87\text{ m}^2$

27. $\sqrt{81} = 9$

28. $\sqrt{64} = 8$

29. $\sqrt{121} = 11$

30. $\sqrt{169} - \sqrt{100} = 13 - 10$
$= 3$

31. $\sqrt{25} + \sqrt{9} + \sqrt{49} = 5 + 3 + 7$
$= 15$

32. $\sqrt{35} \approx 5.916$

33. $\sqrt{48} \approx 6.928$

34. $\sqrt{165} \approx 12.845$

35. $\sqrt{180} \approx 13.416$

36. $\text{hypotenuse} = \sqrt{3^2 + 4^2}$
$= \sqrt{9+16}$
$= \sqrt{25}$
$= 5 \text{ km}$

37. $\text{hypotenuse} = \sqrt{13^2 - 12^2}$
$= \sqrt{169-144}$
$= \sqrt{25}$
$= 5 \text{ yd}$

38. $\text{leg} = \sqrt{20^2 - 18^2}$
$= \sqrt{400-324}$
$= \sqrt{76}$
$\approx 8.72 \text{ cm}$

39. $h = \sqrt{6^2 + 7^2}$
$= \sqrt{36+49}$
$= \sqrt{85}$
$\approx 9.22 \text{ m}$

40. $\text{hypotenuse} = \sqrt{5^2 + 4^2}$
$= \sqrt{25+16}$
$= \sqrt{41}$
$\approx 6.4 \text{ cm}$

41. $\text{hypotenuse} = \sqrt{9^2 + 2^2}$
$= \sqrt{81+4}$
$= \sqrt{85}$
$\approx 9.2 \text{ ft}$

42. $\text{leg} = \sqrt{11^2 - 9^2}$
$= \sqrt{121-81}$
$= \sqrt{40}$
$\approx 6.3 \text{ ft}$

43. $\text{leg} = \sqrt{7^2 - 6^2}$
$= \sqrt{49-36}$
$= \sqrt{13}$
$\approx 3.6 \text{ ft}$

44. $d = 2r$
$= 2(53 \text{ cm})$
$= 106 \text{ cm}$

45. $r = \frac{d}{2}$
$= \frac{126 \text{ cm}}{2}$
$= 63 \text{ cm}$

46. $C = \pi r$
$= 3.14(14 \text{ in.})$
$= 43.96 \text{ in.}$
$\approx 44.0 \text{ in.}$

47. $C = \pi r$
$= 2(3.14)(9 \text{ in.})$
$= 56.52 \text{ in.}$
$\approx 56.5 \text{ in.}$

48. $A = \pi r^2$
$= 3.14(9 \text{ m})^2$
$= 3.14(81 \text{ m})^2$
$= 254.34 \text{ m}^2$
$\approx 254.3 \text{ m}^2$

49. $r = \frac{d}{2} = \frac{8.6 \text{ ft}}{2} = 4.3 \text{ ft}$
$A = \pi r^2$
$= 3.14(4.3 \text{ ft})^2$
$= 3.14(18.49 \text{ ft})^2$
$= 58.06 \text{ ft}^2$

50. $A = \pi r^2 - \pi r^2$
$= 3.14(11\text{ in.})^2 - 3.14(7\text{ in.})^2$
$= 3.14(121\text{ in.}^2) - 3.14(49\text{ in.}^2)$
$= 379.94\text{ in.}^2 - 153.86\text{ in.}^2$
$= 226.08\text{ in.}^2$
$\approx 226.1\text{ in.}^2$

51. $A = \pi r^2 - \pi r^2$
$= 3.14(10\text{ m})^2 - 3.14(7\text{ m})^2$
$= 3.14(100\text{ m}^2) - 3.14(36\text{ m}^2)$
$= 3.14\text{ m}^2 - 113.04\text{ m}^2$
$= 200.96\text{ m}^2$
$\approx 201.0\text{ m}^2$

52. $A = lw + \pi r^2$
$= (24\text{ ft})(10\text{ ft}) + 3.14(5\text{ ft})^2$
$= 240\text{ ft}^2 + 78.5\text{ ft}^2$
$= 318.5\text{ ft}^2$

53. $A = lw - \pi r^2$
$= (20\text{ m})(14\text{ m}) - 3.14(7\text{ m})^2$
$= 280\text{ m}^2 - 153.86\text{ m}^2$
$= 126.14\text{ m}^2$
$\approx 126.1\text{ m}^2$

54. $A = bh - \pi r^2$
$= (12\text{ ft})(10\text{ ft}) - 3.14(2\text{ ft})^2$
$= 120\text{ ft}^2 - 12.56\text{ ft}^3$
$= 107.44\text{ ft}^2$
$\approx 107.4\text{ ft}^2$

55. $A = \dfrac{h(b+B)}{2} + \dfrac{1}{2} \times \pi r^2$
$= \dfrac{5\text{ m}(8\text{ m} + 14\text{ m})}{2} + \dfrac{1}{2} \times (3.14)(4\text{ m})^2$
$= \dfrac{5\text{ m}(22\text{ m})}{2} + \dfrac{1}{2} \times (3.14)(16\text{ m}^2)$
$= 55\text{ m} + 25.12\text{ m}^2$
$= 80.12\text{ m}^2$
$\approx 80.1\text{ m}^2$

56. $V = lwh$
$= (12)(6)(6)$
$= 432\text{ ft}^3$

57. $V = \dfrac{4\pi r^3}{3}$
$= \dfrac{4(3.14)(4.75)^3}{3}$
$= \dfrac{4(3.14)(107.172)}{3}$
$\approx 448.7\text{ in.}^3$

58. $V = \pi r^2 h$
$= 3.14(1.5\text{ ft})^2(3\text{ ft})$
$= 3.14(2.25\text{ ft}^2)(3\text{ ft})$
$\approx 21.2\text{ ft}^3$

59. $V = \pi r^2 h$
$= 3.14(1.5\text{ in.})^2(5\text{ in.})$
$= 3.14(2.25\text{ in.}^2)(5\text{ in.})$
$\approx 35.3\text{ in.}^3$

60. $B = (7\text{ m})(7\text{ m}) = 49\text{ m}^2$
$V = \dfrac{Bh}{3}$
$= \dfrac{(49\text{ m}^2)(15\text{ m})}{3}$
$= 245\text{ m}^3$

61. $V = \frac{\pi r^2 h}{3}$
$= \frac{3.14(20\text{ ft})^2(9\text{ ft})}{3}$
$= \frac{3.14(400\text{ ft}^2)(9\text{ ft})}{3}$
$= 3768\text{ ft}^3$

62. $V = \frac{\pi r^2 h}{3}$
$= \frac{3.14(17\text{ yd})^2(30\text{ yd})}{3}$
$= \frac{3.14(289\text{ yd}^2)(30\text{ yd})}{3}$
$= \frac{27{,}203.8\text{ yd}^3}{3}$
$= 9074.6\text{ yd}^3$

63. $\frac{n}{2} = \frac{40}{3}$
$3n = (2)(45)$
$3n = 90$
$\frac{3n}{3} = \frac{90}{3}$
$n = 30$

30 m

64. $\frac{n}{20} = \frac{6}{36}$
$66n = (120)(6)$
$36n = 120$
$\frac{36n}{36} = \frac{120}{36}$
$n \approx 3.3$

3.3 m

65. Small figure:
$p = 7 + 18 + 7 + 26 = 58\text{ cm}$
$\frac{n}{58} = \frac{108}{18}$
$18n = 108(58)$
$\frac{18n}{18} = \frac{6264}{18}$
$n = 348$

348 cm

66. Small figure:
$p = 13 + 19 + 12 + 26 = 70\text{ ft}$
$\frac{n}{70} = \frac{32.5}{13}$
$13n = 70(32.5)$
$\frac{13n}{13} = \frac{2275}{13}$
$n = 175$

175 ft

67. $3\frac{1}{2} \times 3\frac{1}{2} = \frac{7}{2} \times \frac{7}{2} = \frac{49}{4}$
$\frac{n}{\frac{49}{4}} = \frac{12}{1}$
$n = \left(\frac{49}{4}\right)(12)$
$= 147$

147 yd^2

68. $V = \frac{\pi r^2 h}{3}$
$= \frac{3.14(9\text{ in})^2(24\text{ in})}{3}$
$= \frac{3.14(81\text{ in}^2)(24\text{ in})}{3}$
$\approx 2034.7\text{ in}^3$
$W = 2034.7\text{ in}^3 \times \frac{16\text{ g}}{1\text{ in}^3}$
$= 32{,}555.2\text{ g}$

69. $A = lw - lw$
$= (14\text{ yd})(8\text{ yd}) - (4\text{ yd})(5\text{ yd})$
$= 112\text{ yd}^2 - 20\text{ yd}^2$
$= 92\text{ yd}^2$

$\text{Cost} = 92\text{ yd}^2 \times \frac{\$8}{\text{yd}^2}$
$= \$736$

70. a. $\text{Trip} = 32\text{ km} + 18\text{ km}$
$= 50\text{ km}$

$\text{Speed} = \frac{50\text{ km}}{0.5\text{ hr}}$
$= 100\text{ km/hr}$

b. $\text{Trip} = 26\text{ km} + 14\text{ km} + 16\text{ km}$
$= 56\text{ km}$

$\text{Speed} = \frac{56\text{ km}}{0.8\text{ hr}}$
$= 70\text{ km/hr}$

c. through Ipswich

71. a. $V = \pi r^2 h + \frac{1}{2} \times \frac{4\pi r^3}{3}$
$= 3.14(9\text{ ft})^2(80\text{ ft}) + \frac{1}{2} \times \frac{4(3.14)(9\text{ ft})^3}{3}$
$= 3.14(81\text{ ft}^2)(80\text{ ft}) + \frac{1}{2} \times \frac{4(3.14)(729\text{ ft}^3)}{3}$
$= 20{,}347.2\text{ft}^3 + \frac{1}{2} \times 3052.08\text{ ft}^3$
$= 21{,}873.24\text{ ft}^3$
$\approx 21{,}873.2\text{ ft}^3$

b. $B = 21{,}873.2\text{ ft}^3 \times \frac{0.8\text{ bushel}}{1\text{ ft}^3}$
$\approx 17{,}498.6\text{ bushels}$

72. $2.757\text{ billion} \times 1.244\text{ ft}^3$
$= 3.429708\text{ billion ft}^3$
$= 3{,}429{,}708{,}000\text{ ft}^3$

73. $h = \frac{V}{lw}$
$= \frac{3{,}429{,}108{,}000\text{ ft}^3}{(10{,}000\text{ ft})(20{,}000\text{ ft})}$
$= \frac{3{,}429{,}108{,}000\text{ ft}^3}{200{,}000{,}000\text{ ft}^2}$
$\approx 17.1\text{ ft}$

74. $V = lwh$
$= (2.25)(4)(2)$
$= 18\text{ ft}^3$

$18\text{ ft}^3 \times \frac{62\text{ lb}}{\text{ft}^3} = 1116\text{ lb}$

$1116\text{ lb} \times \frac{1\text{ gal}}{8.6\text{ lb}} \approx 130\text{ gal}$

75. $2\text{ ft} \times \frac{12\text{ in}}{1\text{ ft}} = 24\text{ in.}$

$4\text{ ft} \times \frac{12\text{ in}}{1\text{ ft}} = 48\text{ in.}$

$V = lwh$
$= 24(48)(1.5)$
$= 1728\text{ in.}^3$

76. $C = 2\pi r$
$= 2(3.14)(30\text{ ft})$
$= 188.4\text{ ft}$

$5(188.4) = 942\text{ ft}$

77. $A = lw + \pi r^2,\ r = \frac{d}{2} = \frac{16\text{ yd}}{2} = 8\text{ yd}$
$= (16\text{ yd})(20\text{ yd}) + 3.14(8\text{ yd})^2$
$= 320\text{ yd}^2 + 200.96\text{ yd}^2$
$= 520.96\text{ yd}^2$
$\approx 521.0\text{ yd}^2$

78. $A \approx 521.0 \text{ yd}^2$

$\text{Cost} = 521 \text{ yd}^2 \times \frac{\$50}{1 \text{ yd}^2}$

$= \$26,050$

79. $\text{leg} = \sqrt{33^2 - 30^2}$

$= \sqrt{1089 - 900}$

$= \sqrt{189}$

$\approx 13.7 \text{ ft}$

80. $r = \frac{d}{2} = \frac{90 \text{ m}}{2} = 45 \text{ m}$

$V = \frac{4\pi r^2}{3}$

$= \frac{4(3.14)(45 \text{ m})^3}{2}$

$= \frac{4(3.14)(91,125 \text{ m}^3)}{3}$

$= 381,510 \text{ m}^3$

81. $18 \text{ in} \times \frac{1 \text{ ft}}{12 \text{ in}} = \frac{3}{2} \text{ ft}$

$r = \frac{d}{2} = \frac{\frac{3}{2}}{2} = \frac{3}{4} \text{ ft} = 0.75 \text{ ft}$

$V = \pi r^2 h$

$= 3.14(0.75 \text{ ft})^2(5 \text{ ft})$

$= 3.14(0.5625 \text{ ft}^2)(5 \text{ ft})$

$= 8.83125 \text{ ft}^3$

$\approx 8.8 \text{ ft}^3$

82. $V \approx 8.8 \text{ ft}^3$

$\text{gallons} = 8.8 \text{ ft}^3 \times \frac{7.5 \text{ gal}}{1 \text{ ft}^3}$

$\approx 66 \text{ gal}$

83. $A = \frac{(35 \text{ ft})(45 \text{ ft} + 50 \text{ ft})}{2}$

$= \frac{(35 \text{ ft})(95 \text{ ft})}{2}$

$= \frac{3325 \text{ ft}^2}{2}$

$= 1662.5 \text{ ft}^2$

84. $A = 1662.5 \text{ ft}^2$

$\text{Cost} = 1662.5 \text{ ft}^2 \times \frac{\$0.50}{1 \text{ ft}^2} \times 3 \text{ times/yr}$

$= \$2493.75$

How Am I Doing? Chapter 7 Test

1. $\angle b = \angle a = 52^\circ$

$\angle c = 180^\circ - \angle b = 180^\circ - 52^\circ = 128^\circ$

$\angle e = \angle c = 128^\circ$

2. $P = 2(9 \text{ yd}) + 2(11 \text{ yd})$

$= 18 \text{ yd} + 22 \text{ yd}$

$= 40 \text{ yd}$

3. $P = 4(6.3 \text{ ft})$

$= 25.2 \text{ ft}$

4. $P = 2(6.5 \text{ m}) + 2(3.5 \text{ m})$

$= 13 \text{ m} + 7 \text{ m}$

$= 20 \text{ m}$

5. $P = 2(13 \text{ m}) + 22 \text{ m} + 32 \text{ m}$

$= 26 \text{ m} + 22 \text{ m} + 32 \text{ m}$

$= 80 \text{ m}$

6. $P = 58.6 \text{ m} + 32.9 \text{ m} + 45.4 \text{ m}$

$= 137 \text{ m}$

7. $A = (10 \text{ yd})(18 \text{ yd})$

$= 180 \text{ yd}^2$

8. $A = (10.2 \text{ m})^2$

$= 104.04 \text{ m}^2$

$\approx 104.0 \text{ m}^2$

9. $A = (13 \text{ m})(6 \text{ m})$

$= 78 \text{ m}^2$

10. $A = \frac{(9\text{ m})(7\text{ m} + 25\text{ m})}{2}$
$= \frac{(9\text{ m})(32\text{ m})}{2}$
$= \frac{288\text{ m}^2}{2}$
$= 144\text{ m}^2$

11. $A = \frac{(4\text{ cm})(6\text{ cm})}{2}$
$= \frac{24\text{ cm}^2}{2}$
$= 12\text{ cm}^2$

12. $\sqrt{144} = 12$

13. $\sqrt{169} = 13$

14. $90^\circ - 63^\circ = 27^\circ$

15. $180^\circ - 107^\circ = 73^\circ$

16. $180^\circ - (12.5^\circ + 83.5^\circ) = 180^\circ - 96^\circ$
$= 84^\circ$

17. $\sqrt{54} \approx 7.348$

18. $\sqrt{135} \approx 11.619$

19. $\text{hypotenuse} = \sqrt{7^2 + 5^2}$
$= \sqrt{49 + 25}$
$= \sqrt{74}$
$= 8.602$

20. $\text{leg} = \sqrt{26^2 - 24^2}$
$= \sqrt{676 - 576}$
$= \sqrt{100}$
$= 10$

21. $\text{hypotenuse} = \sqrt{5^2 + 3^2}$
$= \sqrt{25 + 9}$
$= \sqrt{34}$
$\approx 5.83\text{ cm}$

22. $\text{hypotenuse} = \sqrt{15^2 - 12^2}$
$= \sqrt{225 - 144}$
$= \sqrt{81}$
$= 9\text{ ft}$

23. $r = \frac{d}{2} = \frac{18}{2} = 9\text{ ft}$
$C = 2\pi r = 2(3.14)(9) \approx 56.52\text{ ft}$

24. $r = \frac{d}{2} = \frac{12}{2} = 6\text{ ft}$
$A = \pi r^2 = 3.14(6)^2 = 3.14(36) = 113.04\text{ ft}^2$

25. $A = bh - \pi r^2$
$= (15\text{ in.})(8\text{ in.}) - (3.14)(2\text{ in.})^2$
$= 120\text{ in.}^2 - 12.56\text{ in.}^2$
$= 107.44\text{ in.}^2$
$\approx 107.4\text{ in.}^2$

26. $A = \frac{h(b+B)}{2} + \frac{1}{2} \times \pi r^2$
$= \frac{(7\text{ in.})(10\text{ in.} + 20\text{ in.})}{2} + \frac{1}{2} \times (3.14)(5\text{ in.})^2$
$= \frac{(7\text{ in.})(30\text{ in.})}{2} + \frac{1}{2} \times (3.14)(25\text{ in.}^2)$
$= 105\text{ in.}^2 + 39.25\text{ in.}^2$
$= 144.25\text{ in.}^2$
$\approx 144.3\text{ in.}^2$

27. $V = lwh$
$= 3.5(20)(20)$
$= 700\text{ m}^3$

28. $V = \frac{\pi r^2 h}{3}$

$= \frac{3.14(8\text{ m})^2(12\text{ m})}{3}$

$= \frac{3.14(64\text{ m}^2)(12\text{ m})}{3}$

$= 803.84\text{ m}^3$

$\approx 803.8\text{ m}^3$

29. $V = \frac{4\pi r^3}{3}$

$= \frac{4(3.14)(3\text{ m})^3}{3}$

$= \frac{4(3.14)(27\text{ m}^3)}{3}$

$= 113.04\text{ m}^3$

$\approx 113.0\text{ m}^3$

30. $V = \pi r^2 h$

$= 3.14(9\text{ ft})^2(2\text{ ft})$

$= 3.14(81\text{ ft}^2)(2\text{ ft})$

$= 508.68\text{ ft}^3$

$\approx 508.7\text{ ft}^3$

31. $B = (4\text{ m})(3\text{ m}) = 12\text{ m}^2$

$V = \frac{Bh}{3}$

$= \frac{(12\text{ m}^2)(14\text{ m})}{3}$

$= 56\text{ m}^3$

32. $\frac{n}{18} = \frac{413}{5}$

$5n = 18(13)$

$\frac{5n}{5} = \frac{234}{5}$

$n = 46.8\text{ m}$

33. $\frac{n}{7} = \frac{60}{10}$

$10n = 7(60)$

$\frac{10n}{10} = \frac{420}{10}$

$n = 42\text{ ft}$

34. $r = \frac{d}{2} = 20\text{ yd}$

$A = lw - \pi r^2$

$= (130\text{ yd})(40\text{ yd}) + 3.14(20\text{ yd})^2$

$= 5200\text{ yd}^2 + 1256\text{ yd}^2$

$= 6456\text{ yd}^2$

35. $A = 6456\text{ yd}^2$

$\text{Cost} = 6456\text{ yd}^2 \times \frac{\$0.40}{1\text{ yd}^2}$

$= \$2582.40$

Cumulative Test for Chapters 1 - 7

1.

$$\begin{array}{r} 126{,}350 \\ 278{,}120 \\ +\ 531{,}290 \\ \hline 935{,}760 \end{array}$$

2.

$$\begin{array}{r} 163 \\ \times\ 205 \\ \hline 815 \\ 3260 \\ \hline 33{,}415 \end{array}$$

3. $\frac{17}{18} - \frac{11}{12} = \frac{34}{36} - \frac{33}{36} = \frac{1}{36}$

4. $\frac{3}{7} \div 2\frac{1}{4} = \frac{3}{7} \div \frac{9}{4}$

$= \frac{3}{7} \times \frac{4}{9}$

$= \frac{4}{21}$

5. $56.1279 \approx 56.13$

6. $$\begin{array}{r} 9.034 \\ \times \quad 0.8 \\ \hline 7.2272 \end{array}$$

7. $2.634\times 10^2 = 2.634\times 100$
$= 263.4$

8. $$\begin{array}{r} 83 \\ 0.021\overline{)1.743} \\ \underline{168} \\ 63 \\ \underline{63} \\ 0 \end{array}$$

9. $\frac{3}{n}=\frac{2}{18}$
$3\times 8 = n\times 2$
$54 = n\times 2$
$\frac{52}{2}=\frac{n\times 2}{2}$
$27 = n$

10. $\frac{7}{100}=\frac{56}{n}$
$7\times n = 100\times 56$
$\frac{7\times n}{7}=\frac{5600}{7}$
$n = 800$ students

11. $\frac{18}{24} = 0.75 = 75\%$

12. 0.8% of what number is 16?
$0.8\%\times n = 16$
$\frac{0.008\times n}{0.008}=\frac{16}{0.008}$
$n = 2000$

13. What is 15% of 120?
$n = 15\%\times 120$
$= 0.15\times 20$
$= 18$

14. $586 \text{ cm}\times\frac{1 \text{ m}}{100 \text{ cm}} = 5.86 \text{ cm}$

15. $42 \text{ yd}\times\frac{36 \text{ in.}}{1 \text{ yd}} = 1512 \text{ in.}$

16. $88 \text{ km}\times\frac{0.62 \text{ mi}}{1 \text{ km}} = 54.56 \text{ mi}$

17. $P = 2(17 \text{ m}) + 2(8 \text{ m})$
$= 34 \text{ m} + 16 \text{ m}$
$= 50 \text{ m}$

18. $P = 86 \text{ cm} + 13 \text{ cm} + 96 \text{ cm} + 13 \text{ cm}$
$= 208 \text{ cm}$

19. $C = \pi d$
$= 3.14(18 \text{ yd})$
$= 56.52 \text{ yd}$
$\approx 56.5 \text{ yd}$

20. $A = \frac{bh}{2}$
$= \frac{(1.3 \text{ cm})(2.4 \text{ cm})}{2}$
$= \frac{2.88 \text{ cm}^2}{2}$
$= 1.44 \text{ cm}^2$
$\approx 1.4 \text{ cm}^2$

21. $A = \frac{h(b+B)}{2}$
$= \frac{(18 \text{ m})(26 \text{ m}+34 \text{ m})}{2}$
$= \frac{(18 \text{ m})(60 \text{ m})}{2}$
$= 540 \text{ m}^2$

22. $A = lw + bh$
$= (12 \text{ m})(12 \text{ m}) + (12 \text{ m})(4 \text{ m})$
$= 144 \text{ m}^2 + 48 \text{ m}^2$
$= 192 \text{ m}^2$

23. $A = lw - lw$
$= (35 \text{ yd})(20 \text{ yd}) - (6 \text{ yd})(6 \text{ yd})$
$= 700 \text{ yd}^2 - 36 \text{ yd}^2$
$= 664 \text{ yd}^2$

24. $A = \pi r^2$
$= 3.14(4 \text{ m})^2$
$= 3.14(16 \text{ m})^2$
$= 50.24 \text{ m}^2$
$\approx 50.2 \text{ m}^2$

25. $V = 2\pi rh$
$= 2(3.14)(2)(7)$
$\approx 87.9 \text{ m}^3$

26. $r = \frac{d}{2} = \frac{4}{2} = 2 \text{ cm}$
$V = \frac{4\pi r^3}{3}$
$= \frac{4(3.14)(2)^3}{3}$
$= \frac{4(3.14)(8)}{3}$
$\approx 33.5 \text{ cm}^3$

27. $B = (14 \text{ cm})(21 \text{ cm}) = 294 \text{ cm}^2$
$V = \frac{Bh}{3}$
$= \frac{(294 \text{ cm}^2)(32 \text{ m})}{3}$
$= 3136 \text{ cm}^3$

28. $V = \frac{\pi r^2 h}{3}$
$= \frac{3.14(12 \text{ m})^2 (18 \text{ m})}{3}$
$= \frac{3.14(144 \text{ m}^2)(18 \text{ m})}{3}$
$= 2712.96 \text{ m}^3$
$\approx 2713.0 \text{ m}^3$

29. $\frac{n}{9} = \frac{30}{7}$
$7n = 9(30)$
$\frac{7n}{7} = \frac{270}{7}$
$n = 38.6 \text{ m}$

30. $\frac{n}{11} = \frac{1.5}{4}$
$4n = (11)(1.5)$
$4n = 16.5$
$\frac{4n}{4} = \frac{16.5}{4}$
$n = 4.125$
$n \approx 4.1$

4.1 ft

31. a. $A = (14 \text{ yd})(6 \text{ yd}) + (5 \text{ yd})(5 \text{ yd})$
$= \quad + \frac{(5 \text{ yd})(6 \text{ yd})}{2}$
$= 84 \text{ yd}^2 + 25 \text{ yd}^2 + 15 \text{ yd}^2$
$= 124 \text{ yd}^2$

b. $\text{Cost} = 124 \text{ yd}^2 \times \frac{\$8}{1 \text{ yd}^2}$
$= \$992$

32. $\sqrt{144} + \sqrt{81} = 12 + 9 = 21$

33. $\sqrt{57} \approx 7.550$

34. $\text{hypotenuse} = \sqrt{10^2 + 3^2}$
$= \sqrt{100 + 9}$
$= \sqrt{109}$
$\approx 10.440 \text{ in.}$

35. $\text{leg} = \sqrt{7^2 - 5^2}$
$= \sqrt{49 - 25}$
$= \sqrt{24}$
$\approx 4.899 \text{ m}$

36. $\text{hypotenuse} = \sqrt{12^2 + 7^2}$
$= \sqrt{144 + 49}$
$= \sqrt{193}$
$\approx 13.9 \text{ mi}$

37. $20 \text{ ft} \times \frac{12 \text{ in.}}{1 \text{ ft}} = 240 \text{ in.}$

$\text{Number needed} = 240 \div 7\frac{1}{2}$
$= 240 \div \frac{15}{2}$
$= 240 \times \frac{2}{15}$
$= 32 \text{ paintbrushes}$

Chapter 8

8.1 Exercises

1. You multiply $25\% \times 4000$, which is $0.25 \times 4000 = 1000$ students.

3. You would divide the circle into quarters by drawing two perpendicular lines. Shade in one-quarter of the circle. Label this with the title "within five miles = 1000."

5. rent

7. $200

9. $650 + $150 = $800

11. $\frac{\$650}{\$200} = \frac{650 \div 50}{200 \div 50} = \frac{13}{4}$

13. $\frac{\$1000}{\$2700} = \frac{1000 \div 100}{2700 \div 100} = \frac{10}{27}$

15. 40 years old or younger but older than 20

17. 78 million

19. 125 million

21. $\frac{82}{78} = \frac{41}{39}$

23. $\frac{125}{163}$

25. $11\% + 8\% = 19\%$

27. reasonable prices and great food

29. Difference $= 56\% - 22\% = 34\%$

$$n = 34\% \times 1010$$
$$= 0.43 \times 1010$$
$$= 343 \text{ people}$$

31. $n = 35\%$ of $6,300,000,000$

$$= 0.35 \times 6,300,000,000$$
$$= 2,205,000,000 \text{ Christians}$$

33. $19\% + 21\% = 40\%$

35. $100\% - 21\% = 79\%$

37. 68,000,000 is what percent of 6,300,000,000?

$$n \times 6,300,000,000 = 68,000,000$$
$$n = \frac{68,000,000}{6,300,000,000}$$
$$n = 0.011$$
$$n = 1.1\%$$

Cumulative Review

39. $A = \frac{bh}{2}$

$$= \frac{12 \times 20}{2}$$
$$= 120 \text{ ft}^2$$

41. $A = 2lw + 2lw$

$$= 2(7 \text{ yd})(12 \text{ yd}) + 2(7 \text{ yd})(20 \text{ yd})$$
$$2(84 \text{ yd}^2) + 2(140 \text{ yd}^2)$$
$$= 168 \text{ yd}^2 + 280 \text{ yd}^2$$
$$= 448 \text{ yd}^2$$

$$448 \text{ yd}^2 \times \frac{1 \text{ gal}}{28 \text{ yd}^2} = 16 \text{ gal}$$

16 gallons of paint

8.2 Exercises

1. 21 million people

3. 14 million people

5. 1960-1970

7. 22 quadrillion Btu

9. 1970

11. 18 - 14 = 4 quadrillion Btu

13. 22 - 16 = 6 quadrillion Btu

15. From 1975 to 1980 and from 1985 to 1990 with an increase of 3 quadrillion Btu.

17. $24-18=6$ quadrillion Btu
For 2020:
$24+6=30$ quadrillion Btu

19. $1,000,000

21. 1993 to 1995

23. $\text{Increase} = 2,400,000-2,100,000$
$= 300,000$

$\text{Per year} = \dfrac{300,000}{2} = 150,000$

$\text{Future: } 12\text{ yr} \times \dfrac{\$150,000}{y}$
$= \$1,800,000$

$2015: 2,400,000+1,800,000$
$= \$4,200,000$

25. 2.5 in. of rainfall

27. October, November, and December

29. 4.0 - 2.5 = 1.5 in.

31.

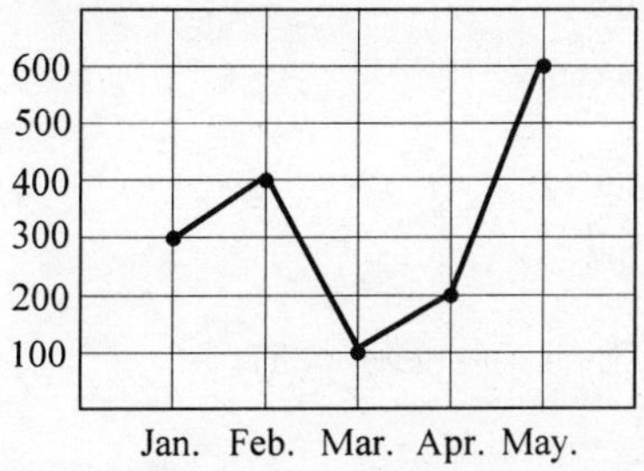

Cumulative Review

33. $8\times 0.5+2.5-5(0.5)$
$=4+2.5-2.5$
$=4$

35. $\dfrac{1}{5}\times\left(\dfrac{1}{5}-\dfrac{1}{6}\right)\times\dfrac{2}{3}=\dfrac{1}{5}+\left(\dfrac{6}{30}-\dfrac{5}{30}\right)\times\dfrac{2}{3}$
$=\dfrac{1}{5}+\dfrac{1}{30}\times\dfrac{2}{3}$

How Am I Doing? Sections 8.1-8.2

1. 14%

2. 2 people

3. 16% + 14% + 6% + 3% = 39%

4. $n=3\%(109,297,000)$
$n=0.03(109,297,000)$
$n=3,278,910$ households

5. $26\%+35\%=61\%$
$n=61\%(109,297,000)$
$=0.61(109,297,000)$
$=66,671,170$ households

6. 300 housing starts

7. 450 housing starts

8. During the second quarter of 2002

9. During the third quarter of 2003

10. 540 - 240 = 300 more starts

11. 420 - 370 = 50 fewer starts

12. August and December

13. December

14. November

15. a. 20,000 sets
b. 30,000 sets

8.3 Exercises

1. The horizontal label for each item in a bar graph is usually a single number or a word title. For the histogram it is a class interval. The vertical bars have a space between them in the bar graph. For the histogram the vertical bars join each other.

3. A class frequency is the number of times a score occurs in a particular class interval.

5. 150 cities

7. 10 cities

9. 20 + 10 = 30 cities

11. 120 + 150 = 270 cities

13. 8,000 books

15. books costing \$5.00 - \$7.99

17.
$$\begin{array}{r} 3,000 \\ 8,000 \\ +\ 17,000 \\ \hline 28,000 \text{ books} \end{array}$$

19.
$$\begin{array}{r} 17,000 \\ 10,000 \\ 12,000 \\ +\ 13,000 \\ \hline 52,000 \text{ books} \end{array}$$

21.
$$\begin{array}{r} 13,000 \\ +\ 7,000 \\ \hline 20,000 \text{ books} \end{array}$$

$$\text{percent} = \frac{20,000}{70,000} \approx 0.286 = 28.6\%$$

23. Tally: 111
Frequency: 3

25. Tally: ~~1111~~1
Frequency: 6

27. Tally: 111
Frequency: 3

29. Tally: 11
Frequency: 2

31.

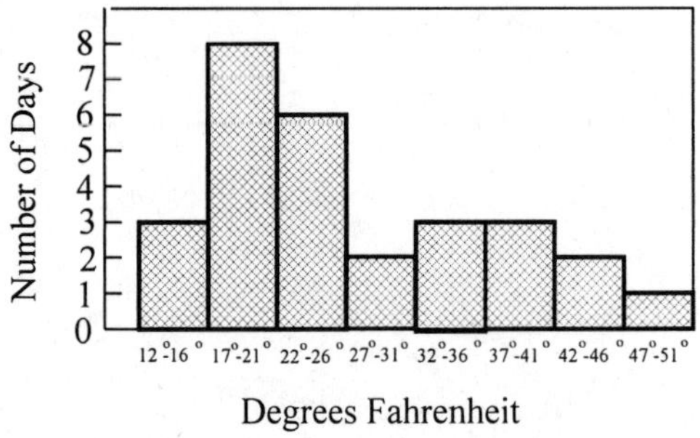

33. 3 + 8 + 6 = 17 days

35. Tally: 111
Frequency: 3

37. Tally: 11
Frequency: 2

39. Tally: 1
Frequency: 1

41. Tally: 11
Frequency: 2

43. 6 + 3 + 3 = 12 prescriptions

Cumulative Review

45. $\dfrac{126}{n} = \dfrac{36}{17}$

$125 \times 17 = n \times 36$

$2142 = n \times 36$

$\dfrac{2142}{36} = \dfrac{n \times 36}{36}$

$59.5 = n$

47. $\dfrac{n}{36} = \dfrac{5}{12}$

$12n = 36 \times 5$

$\dfrac{12n}{12} = \dfrac{180}{12}$

$n = 15$

8.4 Exercises

1. The median of a set of numbers when they are arranged in order from smallest to largest is that value that has the same number of values above it as below it.
The mean of a set of values is the sum of the values divided by the number of values. The mean is most likely to be not typical of the value you would expect if there are many extremely how values or many extremely high values. The median is more likely to be typical of the value you would expect.

3. $\text{Mean} = \dfrac{30+29+28+35+34+37+31}{7}$

$= \dfrac{224}{7}$

$= 32$

5. $\text{Mean} = \dfrac{6+2+3+3.5+2.5+1+0}{7}$

$= \dfrac{18}{7}$

≈ 2.6 hours

7. $\text{Batting Average} = \dfrac{\text{number of hits}}{\text{Times at bat}}$

$= \dfrac{0+2+3+2+2}{5+4+6+5+4}$

$= \dfrac{9}{24}$

$= 0.375$

9. Mean

$= \dfrac{67{,}000+86{,}000+107{,}000+134{,}000+152{,}000}{5}$

$= \dfrac{546{,}000}{5}$

$= 109{,}200$

11. Avg miles/gallon

$= \dfrac{\text{miles driven}}{\text{gallons used}}$

$= \dfrac{276+350+391+336}{12+14+17+14}$

$= \dfrac{1353}{57}$

≈ 23.7

23.7 mi/gal

13. 126, 180, 195, 229, 232

Median = 195

15. 11.6, 11.9, 11.9, 12.1, 12.4, 12.5

$\text{Median} = \dfrac{11.9+12.1}{2} = 12$

17. \$11,600, \$15,700, \$17,000, \$23,500 \$26,700, \$31,500

$\text{Median} = \dfrac{\$17{,}000+\$23{,}0500}{2}$

$= \$20{,}250$

19. 10,12, 18, 21, 25, 28, 31

Median = 21 tables

21. \$97, \$109, \$185, \$207, \$218, \$330, \$420

Median = \$207

23. 1.8, 1.9, 2.0, 2.4, 3.1, 3.1, 3.7

$$\text{Median} = \frac{2.0+2.4}{2} = 2.2$$

25. 72,000, 90,000, 101,000, 157,000

$$\text{Median} = \frac{90,000+101,000}{2} = 95,500$$

27.
$$\begin{array}{r} \$30,000 \\ 74,500 \\ 47,890 \\ 89,000 \\ 57,645 \\ 78,090 \\ 110,370 \\ +\ 65,800 \\ \hline \$553,295 \end{array}$$

$$\text{Mean} = \frac{\$553,295}{8} = \$69,161.88$$

29. 1987, 2576, 3700, 4700, 5000, 7200
8764, 9365

$$\text{Median} = \frac{3700+5000}{2} = 4850$$

31.
$$\begin{array}{r} 120.50 \\ 66.74 \\ 80.95 \\ 210.52 \\ +\ 45.00 \\ \hline 523.71 \end{array}$$

$$\text{Mean} = \frac{523.71}{5} = \$104.74$$

45, 66.74, 80.95, 120.5, 210.52
Median = \$80.95

33. The mode is 60 which occurs twice.

35. The modes are 121 and 150 which both occur twice.

37. 249, 649, 269, 259, 269, 249, 269
Mode = \$269

39. Mean

$$= \frac{869+992+482+791+399+855+869}{7} = \frac{5257}{7} = 751$$

399, 482, 791, 855, 869, 992
Median = 855
Mode = 869 since it occurs twice

41. a. $$\text{Mean} = \frac{1500+1700+1650+1300+1440 +1580+1820+1380+2900+6300}{10} = \frac{21,570}{10} = \$2157$$

b. 1300, 1380, 1440, 1500, 1580, 1650, 1700, 1820, 2900, 6300

$$\text{Median} = \frac{1580+1650}{2} = \$1615$$

c. There is no mode

d. The median, because the mean is affected by the high amount \$6300.

43. a. $$\text{Mean} = \frac{23+3+2+3+7+10+11}{7} \approx 8.4 \text{ nights}$$

b. 3, 3, 6, 7, 8, 9, 28
Median = 7 nights

c. Mode = 3 nights

d. The median is the most representative. On three nights she gets more calls than 7. On three nights she get fewer calls than 7. On one night she got 7 calls. The mean is distorted a little because of the very large number of calls on Sunday night. The mode is artificially low because she get so few calls on Monday and Wednesday and it just happened to be the same number, 3.

45. $\text{GPA} = \frac{4(3)+3(4)+2(3)+3(3)}{3+4+3+3}$

$= \frac{12+12+6+9}{3+4+3+3}$

$= \frac{39}{13}$

≈ 3.0

Cumulative Review

47. $A = \frac{bh}{2}$

$= \frac{(7 \text{ in.})(5.5 \text{ in.})}{2}$

$= \frac{38.5 \text{ in.}^2}{2}$

$= 19.25 \text{ in.}^2$

$\approx 19.3 \text{ in.}^2$

49. $A = bh = (5 \text{ ft})(4 \text{ ft}) = 20 \text{ ft}^2$

$\text{Cost} = 20 \text{ ft}^2 \times \frac{\$16.50}{1 \text{ ft}^2} = \$330$

Putting Your Skills to Work

1. Garciaparra: $\frac{198}{658} = 0.301$

Rodriguez: $\frac{181}{607} = 0.298$

Gonzalez: $\frac{122}{536} = 0.228$

Renteria: $\frac{194}{587} = 0.330$

Cabrera: $\frac{186}{626} = 0.297$

AL: Jeter 0.324

NL: Renteria 0.330

2. Jeter: $\frac{217}{482} = 0.450$

Garciaparra: $\frac{345}{658} = 0.524$

Rodriguez: $\frac{364}{607} = 0.600$

Renteria: $\frac{282}{587} = 0.480$

Cabrera: $\frac{288}{626} = 0.460$

AL: Rodriguez 0.600

NL: Renteria 0.480

3. Jeter: $\frac{159+271}{159+271+14} = 0.968$

Rodriguez: $\frac{227+464}{227+464+8} = 0.989$

Gonzalez: $\frac{193+422}{193+422+10} = 0.984$

Renteria: $\frac{191+439}{191+439+16} = 0.975$

Cabrera: $\frac{258+456}{258+456+18} = 0.975$

AL: Rodriguez 0.989

NL: Gonzalez 0.984

4. Answers may vary.

Chapter 8 Review Problems

1. 13 computers

2. 32 computers

3. 43 + 25 = 68 computers

4. 21 + 6 = 27 computers

5. $\frac{13}{21}$

6. $\frac{43}{32}$

7. $\frac{25}{140} \approx 0.179 = 17.9\%$

8. $\frac{32}{140} \approx \frac{32}{140} \approx 0.229 = 22.9\%$

9. $23\% + 25\% = 48\%$

10. 100% - 23% = 77%

11. art

12. business

13. $8\% + 12\% = 20\% = \frac{1}{5}$

art and education

14. $n = 15\%(8000)$
$= 0.15(8000)$
$= 1200$ students

15. $n = 42\%(8000)$
$= 0.42(8000)$
$= 3360$ students

16. $23\% - 12\% = 11\%$
$n = 11\%(8000)$
$= 0.11(8000)$
$= 880$ students

17. 36

18. 26

19. 10-13 years

20. 6-9 years

21. 18 - 6 = 12 glasses

22. 30 - 8 = 22 glasses

23. $\frac{10}{2} = \frac{5}{1}$

24. $\frac{36}{18} = \frac{2}{1}$

25. $31,000

26. $58,000

27. between 1985 and 1990 with a difference of
$\$44,000 - \$33,000 = \$11,000$

28. between 1980 and 1985 with a difference of
$\$24,000 - \$16,000 = \$8,000$

29. $33,000 - $24,000 = $9000

30. $52,000 - $37,000 = $15,000

31. 2000 with a difference of
$\$58,000 - \$41,000 = \$17,000$

32. 1980 with a difference of
$\$23,000 - \$16,000 = \$7,000$

33. $\text{Average} = \frac{61-23}{5} = \frac{38}{5} = 7.6$
$7,600

34. $\text{Average} = \frac{43-16}{5} = \frac{27}{5} = 5.4$
$5,400

35. $\text{Increase} = 37 - 16 = \$21$
$2010:\ 37 + 21 = 58$
$58,000

36. $\text{Difference} = \$58,000 - \$44,000$
$= \$14,000$
$\text{Year } 2005 = \$58,000 + \$14,000$
$= \$72,000$

37. 400 students

38. 500 students

39. 650 students

40. 450 students

41. 300 - 200 = 100 students

42. 500 - 450 = 50 students

43. 1999-2000

44. 2000-2001

45. 45,000 cones

46. 30,000 cones

47. 20,000 - 10,000 = 10,000 cones

48. 60,000 - 30,000 = 30,000 cones

49. 55,000 - 20,000 = 35,000 cones

50. 60,000 - 30,000 = 30,000 cones

51. The sharp drop in the number of ice cream cones purchased from July 2000 to August 2000 is probably directly related to the weather. Since August was cold and rainy, significantly fewer people wanted ice cream during August.

52. The sharp increase in the number of ice cream cones purchased from June 2003 to July 2003 is probably directly related to the weather. Since June was cold and rainy and July was warm and sunny, significantly more people wanted ice cream during July.

53. 14,000,000

54. 18,000,000

55. Between 1996 and 1998

56. Between 1990 and 1992

57. 18,000,000 - 17,000,000 = 1,000,000

58. 33,000,000 - 21,000,000 = 12,000,000

59. 1992

60. 2002

61. $$\begin{aligned}\text{Average} &= \frac{1+2+4+8+3+4}{6}\\ &= \frac{22}{6}\text{ million}\\ &\approx 3{,}666{,}667\end{aligned}$$

62. $\text{Increase} = 18-10 = 8$

$$\begin{aligned}\text{Average} &= \frac{8{,}000{,}000}{6}\\ &\approx 1{,}333{,}333\text{ machines}\end{aligned}$$

63. $21{,}000{,}000 - 18{,}000{,}000 = 3{,}000{,}000$

Year 2004:

$21{,}000{,}000 + 3{,}000{,}000 = 24{,}000{,}000$

64. $33{,}000{,}000 - 17{,}000{,}000 = 16{,}000{,}000$

Year 2006:

$33{,}000{,}000 + 16{,}000{,}000 = 49{,}000{,}000$

65. 65 pairs

66. 10 pairs

67. 55 + 65 + 25 = 145 pairs

68. $$\begin{aligned}\text{Sold} &= 5+20+55+65+25+10\\ &= 180\text{ pairs}\end{aligned}$$

$$p = \frac{180}{200} = 0.9 = 90\%$$

69. Difference = 65 - 20 = 45 pairs

70. $\frac{25}{180} = \frac{5}{36}$

71. Tally: 𝍸 𝍸

Frequency: 10

72. Tally: 𝍸 111

Frequency: 8

73. Tally: 111

Frequency: 3

74. Tally: ~~IIII~~
Frequency: 5

75. Tally: II
Frequency: 2

76.

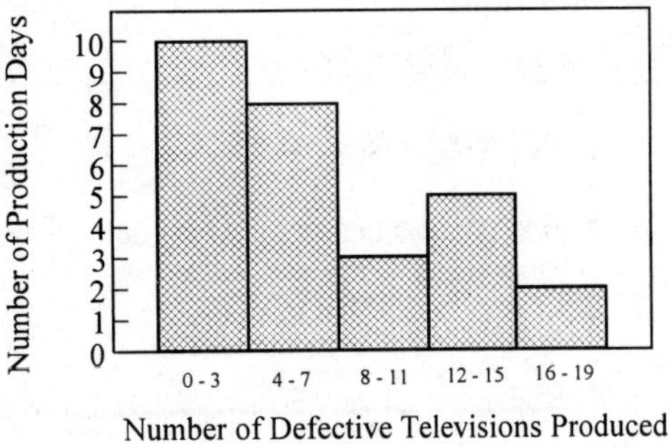

77. 10 + 8 = 18 times

78. $$\text{Mean} = \frac{86+83+88+95+97+100+81}{7} = \frac{630}{7} = 90 = 90^\circ\text{F}$$

79. $$\text{Mean} = \frac{145+162+95+67+43+26}{6} = \$89.67$$

80. $$\text{Mean} = \frac{11,000+19,000+40,000+52,000+23,000}{5} = 29,000 \text{ people}$$

81. $$\text{Mean} = \frac{882+913+1017+1592+1778+1936}{7} = \frac{8118}{6} = 1353 \text{ employees}$$

82. \$21,690, \$28,500, \$29,300, \$35,000
\$37,000, \$38,600, \$43,600, \$45,300

$$\text{Median} = \frac{\$35,000+\$37,000}{2} = \$36,000$$

83. \$98,000, \$120,000, \$126,000, \$135,000
\$139,000, \$144,000, \$150,000, \$154,000
\$156,000, \$170,000

$$\text{Median} = \frac{\$139,000+\$144,000}{2} = \$141,500$$

84. 4, 7, 8, 10, 15, 28, 28, 30, 31, 34, 35, 38, 43, 54, 77, 79

$$\text{Median} = \frac{30+31}{2} = 30.5 \text{ years}$$

Mode = 28 years

85. 3, 9, 13, 14, 15, 15, 16, 18, 19, 21, 24, 25, 26, 28, 31, 36

$$\text{Median} = \frac{18+19}{2} = 18.5 \text{ deliveries}$$

Mode = 15 deliveries

86. The median, because of one low score, 31.

87. The median is better because the means is skewed by the one high data item, 39.

88. a. $\text{Mean} = \frac{2+3+2+4+7+12+5}{7}$

$= 5 \text{ hours}$

b. 2, 2, 3, 4, 5, 7, 12

Median = 4 hours

c. Mode = 2 hours

d. The median is the most representative. On three days she uses the computer more than 4 hours and on three days she uses the computer less than 4 hours. One day she used it exactly 7 hours. The mean is distorted a little because of the very large number of hours on Friday. The mode is artificially low because she happened to use the computers only two hours on Sunday and Tuesday. All other days it was more than this.

How Am I Doing? Chapter 8 Test

1. 37%

2. 21%

3. 6% + 2% + 4% = 12%

4. $30\% \times 300,000$

$= 0.30 \times 300,000$

$= 90,000 \text{ automobiles}$

5. $21\% + 6\% = 27\%$

$27\% \times 300,000 = 0.27 \times 300,000$

$81,000 \text{ automobiles}$

6. $12,000

7. $3000

8. 20,000 - 12,000 = $8000

9. 5000 - 3000 = $2000

10. 14,000 - 3000 = $11,000

11. 20,000 - 5000 = $15,000

12. 20 years

13. 26 years

14. 48 - 36 = 12 years

15. age 35

16. age 65

17. 60,000 televisions

18. 25,000 televisions

19. 15,000 + 5000 = 20,000 televisions

20. 45,000 + 15,000 = 60,000 televisions

21. $10+16+15+12+18+17+14+10$

$+13+20$

$=145$

$\text{Mean} = \frac{145}{10} = 14.5$

22. 10, 10, 12, 13, 14, 15, 16, 17, 18, 20

$\text{Median} = \frac{14+15}{2} = 14.5$

23. Mode = 10

24. Mean or median

Cumulative Test for Chapters 1 - 8

1.

$$\begin{array}{r} 1,376 \\ 2,804 \\ 9,003 \\ +\ 7,642 \\ \hline 20,825 \end{array}$$

2.
$$\begin{array}{r} 2008 \\ \times\ \ 37 \\ \hline 14056 \\ 6024\ \\ \hline 74,296 \end{array}$$

3.
$$\begin{array}{r} 7\tfrac{1}{5} \\ -\ 3\tfrac{3}{8} \\ \hline \end{array} \qquad \begin{array}{r} 6\tfrac{48}{40} \\ -\ 3\tfrac{15}{40} \\ \hline 3\tfrac{33}{40} \end{array}$$

4. $10\frac{3}{4} \div \frac{3}{8} = \frac{43}{4} \times \frac{8}{3} = \frac{86}{3}$ or $28\frac{2}{3}$

5. 1796.4289 rounds to 1796.

6.
$$\begin{array}{r} 200.58 \\ -\ 127.93 \\ \hline 72.65 \end{array}$$

7.
$$\begin{array}{r} 72.23 \\ 0.72\overline{)52.0056} \\ \underline{504} \\ 160 \\ \underline{144} \\ 165 \\ \underline{144} \\ 216 \\ \underline{216} \\ 0 \end{array}$$

8.
$$\begin{aligned} \frac{7}{n} &= \frac{35}{3} \\ 7 \times 3 &= n \times 35 \\ \frac{21}{35} &= \frac{n \times 35}{35} \\ 0.6 &= n \end{aligned}$$

9.
$$\begin{aligned} \frac{n}{26,390} &= \frac{3}{2030} \\ 2030n &= 3 \times 26,390 \\ \frac{2030n}{2030} &= \frac{79,170}{2030} \\ n &= 39 \text{ defects} \end{aligned}$$

10.
$$\begin{aligned} n &= 1.3\% \times 25 \\ n &= 0.013 \times 25 \\ n &= 0.325 \end{aligned}$$

11.
$$\begin{aligned} 20\% \times n &= 12 \\ 0.2n &= 12 \\ \frac{0.2n}{0.2} &= \frac{12}{0.2} \\ n &= 60 \end{aligned}$$

12. 198 cm = 1.98 m (move 2 places left)

13. $18 \text{ yd} \times \frac{3 \text{ ft}}{1 \text{ yd}} = 54 \text{ ft}$

14.
$$\begin{aligned} A &= \pi r^2 \\ &= 3.14(3 \text{ in.})^2 \\ &= 3.14(9 \text{ in.}^2) \\ &\approx 28.3 \text{ in.}^2 \end{aligned}$$

15. $P = 4s = 4(15) = 60$ ft

$A = s^2 = (15)^2 = 225 \text{ ft}^2$

16. 18% + 16% = 34%

17. $12\% + 8\% = 20\%$

$n = 20\% \times 280 = 0.2(280)$

$= 56$ people

18. \$3 million

19.
$$\begin{array}{l} \$7,000,000 \text{ profit in 2nd quarter 2001} \\ \underline{-\$6,000,000 \text{ profit in 2nd quarter 2000}} \\ \$1,000,000 \text{ more profit in 2nd quarter 2001} \end{array}$$

20. 16 in.

21. In 1960 and 1970 the annual rainfall in Weston was greater than Dixville.

22. 8 students

23.

$$\begin{array}{r} 4 \text{ students ages 17-19} \\ 7 \text{ students ages 20-22} \\ + \ 5 \text{ students ages 23-25} \\ \hline 16 \text{ students less than 26 years of age} \end{array}$$

24. $$\text{Mean} = \frac{\begin{array}{c}\$5.00 + \$4.50 + \$3.95 + \$4.90 \\ + \$12.15 + \$4.50 + \$6.00\end{array}}{2}$$

$$= \frac{\$48.00}{8}$$

$$= \$6.00$$

25. \$3.95, \$4.50, \$4.50, \$4.90, \$5.00, \$6.00, \$7.00, \$12.15

$$\text{Median} = \frac{\$4.90 + \$5.00}{2}$$

$$= \$4.95$$

26. Mode = \$4.50

Chapter 9

9.1 Exercises

1. Find the absolute value of each number. Then add those two absolute values. Use the common sign in the answer.

3. $-9 < 2$

5. $-3 > -5$

7. $5 > -2$

9. $-12 < -10$

11. $|7| = 7$

13. $|-16| = 16$

15. $-6 + (-11) = -17$

17. $-4.9 + (-2.1) = -7$

19. $8.9 + 7.6 = 16.5$

21. $\frac{1}{5} + \frac{2}{7} = \frac{7}{35} + \frac{10}{35}$
$= \frac{17}{35}$

23. $-2\frac{1}{2} + \left(-\frac{1}{2}\right) = -2 + \left(-\frac{2}{2}\right) = -3$

25. $14 + (-5) = 9$

27. $-17 + 12 = -5$

29. $-36 + 58 = 22$

31. $-9.3 + 6.05 = -3.25$

33. $\frac{1}{12} + \left(-\frac{3}{4}\right) = \frac{1}{12} + \left(-\frac{9}{12}\right)$
$= -\frac{8}{12}$
$= -\frac{2}{3}$

35. $\frac{7}{9} + \left(-\frac{2}{9}\right) = -\frac{5}{9}$

37. $-18 + (-4) = -22$

39. $1.48 + (-2.2) = -0.72$

41. $-125 + (-238) = -363$

43. $13 + (-9) = 4$

45. $-2\frac{1}{4} + \left(-1\frac{5}{6}\right) = -3 + \left(-\frac{3}{12} - \frac{10}{12}\right)$
$= -3 - \frac{13}{12}$
$= -4\frac{1}{12}$

47. $-7.56 + 13.8 = 6.24$

49. $-5 + \left(-\frac{1}{2}\right) = -\frac{10}{2} + \left(-\frac{1}{2}\right)$
$= -\frac{11}{2}$
$= -5\frac{1}{2}$

51. $-2.05 + 18.1 + (-12.3) = -32.8 + 18.1$
$= -14.7$

53. $11+(-9)+(-10)+8=2+(-10)+8$
$=-8+8$
$=0$

55. $-7+6+(-2)+5+(-3)+(-5)$
$=-1+(-2)+5+(-3)+(-5)$
$=-3+5+(-3)+(-5)$
$=2+(-3)+(-5)$
$=-1+(-5)$
$=-6$

57. $\left(-\frac{1}{5}\right)+\left(-\frac{2}{3}\right)+\left(\frac{4}{25}\right)$
$=\left(-\frac{15}{75}\right)+\left(-\frac{50}{75}\right)+\left(\frac{12}{75}\right)$
$=-\frac{65}{75}+\frac{12}{75}$
$=-\frac{53}{75}$

59. $-\$43,000+(-\$51,000)=-\$94,000$

61. $\$28,000+(-\$19,000)=\$9,000$

63. $-\$35,000+\$17,000+(-\$20,000)$
$=-\$18,000+(-\$20,000)$
$=-\$38,000$

65. $-1^\circ+(-17)^\circ=-18^\circ\text{F}$

67. $-5^\circ+4=-1^\circ\text{F}$

69. $-0.15+0.20+0.21+(-0.24)+(-0.36)$
$=0.41+(-0.75)$
$=-0.34$

71. $-8+13(-6)=5+(-6)$
$=-1$ yard or a loss of 1 yard

73. $-\$28+\$30+(-\$15)=\$2+(-\$15)$
$=-\$13$

75. $\$89.50+(-\$50.00)+(-\$2.50)$
$=\$39.50+(-\$2.50)$
$=\$37.00$

Cumulative Review

77. $v=\frac{4\pi r^3}{3}$
$=\frac{4(3.14)(16\text{ in.})^3}{3}$
$\approx 904.3\text{ ft}^3$

79. 57, 59, 60, 60, 61, 62
Median $=\frac{60+60}{2}=60$

9.2 Exercises

1. $-9-(-3)=-9+3=-6$

3. $-12-(-7)=-12+7=-5$

5. $3-9=3+(-9)=-6$

7. $-14-3=-14+(-3)=-17$

9. $-12-(-10)=-12+10=-2$

11. $46-(-39)=46+39=85$

13. $12-30=12+(-30)=-18$

15. $-12-(-15)=-12+15=3$

17. $150-210=150+(-210)=-60$

19. $300-(-256)=300+256=556$

21. $-2.5-4.2=-2.5+(-4.2)=-6.7$

23. $4.2-10.7=4.2+(-10.7)=-6.5$

25. $-10.9-(-2.3)=-10.9+2.3=-8.6$

27. $20.23-(-12.71)=20.33+12.71=32.94$

29. $\frac{1}{4}-\left(-\frac{3}{4}\right)=\frac{1}{4}+\frac{3}{4}=\frac{4}{4}=1$

31. $$\begin{aligned}-\frac{5}{6}-\frac{1}{3}&=-\frac{5}{6}+\left(-\frac{1}{3}\right)\\&=-\frac{5}{6}+\left(-\frac{2}{6}\right)\\&=-\frac{7}{6}\\&=-1\frac{1}{6}\end{aligned}$$

33. $$\begin{aligned}-7\frac{2}{5}-\left(-2\frac{1}{3}\right)&=-7\frac{6}{15}+2\frac{5}{15}\\&=-5\frac{1}{15}\end{aligned}$$

35. $$\begin{aligned}\frac{5}{9}-\frac{2}{7}&=\frac{5}{9}+\left(-\frac{2}{7}\right)\\&=\frac{35}{63}+\left(-\frac{18}{63}\right)\\&=\frac{17}{63}\end{aligned}$$

37. $$\begin{aligned}2-(-8)+5&=2+8+5\\&=10+5\\&=15\end{aligned}$$

39. $$\begin{aligned}-5-6-(-11)&=-5+(-6)+5\\&=-15+5\\&=-10\end{aligned}$$

41. $21-(-15)-(-10)=21+15+10=46$

43. $$\begin{aligned}-16-(-6)-12&=-16+6+(-12)\\&=-10+(-12)\\&=-22\end{aligned}$$

45. $$\begin{aligned}9-3-2-6&=9+(-3)+(-2)+(-6)\\&=6+(-2)+(-6)\\&=4+(-6)\\&=-2\end{aligned}$$

47. $$\begin{aligned}&-2.4-7.1+1.3-(-2.8)\\&=-2.4+(-7.1)+1.3+2.8\\&=-9.5+4.1\\&=-5.4\end{aligned}$$

49. $14,494-(-282)=14,776$ ft

51. $23^\circ-(-19^\circ)=23^\circ+19^\circ=42^\circ$F

53. $-29^\circ+16^\circ=-13^\circ$F

55. $18,700+(-34,700)=-\$16,000$

57. $-6,300+43,600=\$37,300$

59. $$\begin{aligned}&\$15\frac{1}{2}-\$1\frac{1}{2}+\$2\frac{3}{4}-\$3\frac{1}{4}\\&=\$15\frac{1}{2}+\left(-\$1\frac{1}{2}\right)+\$2\frac{3}{4}+\left(-\$3\frac{1}{4}\right)\\&=\$14+\$2\frac{3}{4}+\left(-\$3\frac{1}{4}\right)\\&=\$16\frac{3}{4}+\left(-\$3\frac{1}{4}\right)\\&=\$13\frac{2}{4}\\&=\$13\frac{1}{2}\end{aligned}$$

61. The bank finds that a customer has \$50 in his checking account. However, the bank must remove an erroneous dedit of \$80 from the customer's account. When the bank makes the correction, what will the new balance be?

Cumulative Review

63. $20 \times 2 \div 10 + 4 - 3$
$= 40 \div 10 + 4 - 3$
$= 4 + 4 - 3$
$= 8 + -3$
$= 8 + (-3)$
$= 5$

65. $r = 6 \div 2 = 3$
$A = \pi r^2$
$= 3.14(3 \text{ in.})^2$
$= 28.26$ square inches

9.3 Exercises

1. To multiply two numbers with the same signs, multiply the absolute values. The sign of the result is positive.

3. $(12)(3) = 36$

5. $(-20)(-3) = 60$

7. $(-20)(8) = -160$

9. $3(-22) = -66$

11. $(2.5)(-0.6) = -1.5$

13. $(-12.5)(-2.25) = 28.125$

15. $\left(-\frac{2}{5}\right)\left(\frac{3}{7}\right) = -\frac{6}{35}$

17. $\left(-\frac{4}{12}\right)\left(-\frac{3}{23}\right) = \frac{1}{23}$

19. $-64 \div 8 = -8$

21. $\frac{48}{-6} = -8$

23. $\frac{-120}{-20} = 6$

25. $-25 \div (-5) = 5$

27. $-\frac{4}{9} \div \left(-\frac{16}{27}\right) = -\frac{4}{9} \times \left(-\frac{27}{16}\right)$
$= \frac{3}{4}$

29. $\frac{-\frac{4}{5}}{-\frac{7}{10}} = -\frac{4}{5} \div \left(-\frac{7}{10}\right)$
$= -\frac{4}{5} \times \left(-\frac{10}{7}\right)$
$= \frac{8}{7}$
$= 1\frac{1}{7}$

31. $50.28 \div (-6) = -8.38$

33. $\frac{45.6}{-8} = -5.7$

35. $\frac{-21{,}000}{-700} = 30$

37. $5(-9) = -45$

39. $(-12)(-4) = 48$

41. $\frac{15}{-3} = -5$

43. $-30 \div (-3) = 10$

45. $(-1.4)(2) = -2.8$

47. $0.028 \div (-1.4) = -0.02$

49. $\left(-\frac{3}{5}\right) \div \left(-\frac{5}{7}\right) = \frac{3}{7}$

51. $\left(-\frac{8}{5}\right)\div\left(-\frac{16}{15}\right)=\left(-\frac{8}{5}\right)\times\left(-\frac{15}{16}\right)=\frac{3}{2}$

53. $3(-6)(-4)=-18(-4)=72$

55. $(-8)(4)(-6)=-32(-6)=192$

57. $$\begin{aligned}2(-8)(3)\left(-\frac{1}{3}\right)&=-16(3)\left(-\frac{1}{3}\right)\\&=-48\left(-\frac{1}{3}\right)\\&=16\end{aligned}$$

59. $$\begin{aligned}(-20)(6)(-30)(-5)&=(-120)(-30)(-5)\\&=(3600)(-5)\\&=-18,000\end{aligned}$$

61. $$\begin{aligned}8(-3)(-5)(0)(-2)&=-24(-5)(0)(-2)\\&=120(0)(-2)\\&=0(-2)\\&=0\end{aligned}$$

63. $$\begin{aligned}\left(-\frac{2}{3}\right)\left(-\frac{3}{4}\right)\left(-\frac{5}{6}\right)&=\left(\frac{1}{2}\right)\left(-\frac{5}{6}\right)\\&=-\frac{5}{12}\end{aligned}$$

65. $90(-1.50)+70(0.80)$
$=-135+56$
$=-79$
She lost \$79.

67. $$\frac{-12^\circ+(-14^\circ)+(-3^\circ)+(-1^\circ)+(-10^\circ)+(-23^\circ)+5^\circ+8^\circ}{8}$$
$=-\frac{50^\circ}{8}$
$=-\frac{25^\circ}{4}$
$=-6.25^\circ$

69. $7(-10)=-70$
Dropped 70 feet

71. $17(2)=34$

73. $4(1)+2(-1)=4+(-2)=2$

75. $-8(-1)=8$

77. $$\begin{aligned}2(-1)+(-2)+2(2)&=-2+(-2)+4\\&=-4+4\\&=0\end{aligned}$$
0; at par

79. The mystery number is 0.

81. b is negative

Cumulative Review

83. $$\begin{aligned}A=bh&=(15\text{ in.})(6\text{ in.})\\&=90\text{ square inches}\end{aligned}$$

How Am I Doing? Sections 9.1 - 9.3

1. $-7+(-12)=-19$

2. $-23+19=-4$

3. $7.6+(-3.1)=4.5$

4. $8+(-5)+6+(-9)=14+(-14)=0$

5. $\frac{5}{12}+\left(-\frac{3}{4}\right)=\frac{5}{12}+\left(-\frac{9}{12}\right)=-\frac{4}{12}=-\frac{1}{3}$

6. $-\frac{5}{6}+\left(-\frac{1}{3}\right)=-\frac{5}{6}+\left(-\frac{2}{6}\right)=-\frac{7}{6}$ or $-1\frac{1}{6}$

7. $-2.8+(-4.2)=-7$

8. $-3.7+5.4=1.7$

9. $13-21=13+(-21)=-8$

10. $-26-15=-26+(-15)=-41$

11. $\frac{5}{17}-\left(-\frac{9}{17}\right)=\frac{5}{17}+\frac{9}{17}=\frac{14}{17}$

12. $-19-(-7)=-19+7=-12$

13. $-4.9-(-6.3)=-4.9+6.3=1.4$

14. $2.8-5.6=2.8+(-5.6)=-2.8$

15. $21-(-21)=21+21=42$

16. $\frac{2}{3}-\left(-\frac{3}{5}\right)=\frac{2}{3}+\frac{3}{5}=\frac{10}{15}+\frac{9}{15}=\frac{19}{15}$ or $1\frac{4}{15}$

17. $(-3)(-8)=24$

18. $-48\div(-12)=4$

19. $-72\div 9=-8$

20. $$\begin{aligned}&(5)(-4)(2)(-1)\left(-\frac{1}{4}\right)\\&=-20(2)(-1)\left(-\frac{1}{4}\right)\\&=-40(-1)\left(-\frac{1}{4}\right)\\&=40\left(-\frac{1}{4}\right)\\&=-10\end{aligned}$$

21. $\frac{72}{-3}=-24$

22. $\frac{-\frac{3}{4}}{-\frac{4}{5}}=-\frac{3}{4}\left(-\frac{5}{4}\right)=\frac{15}{16}$

23. $(-8)(-2)(-4)=16(-4)=-64$

24. $120\div(-12)=-10$

25. $18-(-6)=18+6=24$

26. $-7(-3)=21$

27. $28\div(-4)=-7$

28. $$\begin{aligned}&1.6+(-1.8)+(-3.4)\\&=-0.2+(-3.4)\\&=-3.6\end{aligned}$$

29. $2.9-3.5=2.9+(-3.5)=-0.6$

30. $$\begin{aligned}\left(-\frac{1}{3}\right)+\left(-\frac{2}{5}\right)&=\left(-\frac{5}{15}\right)+\left(-\frac{6}{15}\right)\\&=-\frac{11}{15}\end{aligned}$$

31. $\left(-\frac{5}{6}\right)\left(\frac{3}{7}\right)=-\frac{5}{14}$

32. $$\begin{aligned}\frac{3}{4}\div\left(-\frac{5}{6}\right)&=\frac{3}{4}\left(-\frac{6}{5}\right)\\&=-\frac{9}{10}\end{aligned}$$

33. $$\begin{aligned}\text{Average}&=\frac{-12+(-8)+3+7+(-2)}{5}\\&=\frac{-2}{5}\\&=-0.4^\circ\text{F}\end{aligned}$$

9.4 Exercises

1. $-8\div(-4)(3)=2(3)=6$

3. $50\div(-25)(4)=(-2)(4)=-8$

5. $$\begin{aligned}16+32\div(-14)&=16+(-8)\\&=8\end{aligned}$$

7. $$\begin{aligned}&24\div(-3)+16\div(-4)\\&=-8+(-4)\\&=-12\end{aligned}$$

9. $$\begin{aligned}3(-4)+5(-2)-(-3)&=-12+(-10)+3\\&=-22+3\\&=-19\end{aligned}$$

11. $-54 \div 6(9) = -9(9) = -81$

13. $5 - 30 \div 3 = 5 - 10$
$= 5 + (-10)$
$= -5$

15. $36 \div 12(-2) = 3(-2) = -6$

17. $3(-4) + 6(-2) - 3 = -12 + (-12) + (-3)$
$= -24 + (-3)$
$= -27$

19. $11(-6) - 3(12) = -66 - 36 = -66 + (-36) = -102$

21. $16 - 4(8) + 18 \div (-9)$
$= 16 - 32 + (-2)$
$= 16 + (-32) + (-2)$
$= -16 + (-2)$
$= -18$

23. $\frac{8+6-12}{3-6+5} = \frac{2}{2} = 1$

25. $\frac{6(-2)+4}{6-3-5} = \frac{-12+4}{6-3-5} = \frac{-8}{-2} = 4$

27. $\frac{2(8) \div 4 - 5}{-35 \div (-7)} = \frac{16 \div 4 - 5}{5}$
$= \frac{4-5}{5}$
$= -\frac{1}{5}$

29. $\frac{24 \div (-3) - (6-2)}{-5(4)+8} = \frac{24 \div (-3) - 4}{-20+8}$
$= \frac{-8-4}{-12}$
$= \frac{-12}{-12}$
$= 1$

31. $\frac{12 \div 3 + (-2)(2)}{9 - 9 \div (-3)} = \frac{4 + (-4)}{9 - (-3)}$
$= \frac{4 + (-4)}{9 + 3}$
$= \frac{0}{12}$
$= 0$

33. $3(2-6) + 4^2 = 3(-4) + 4^2$
$= 3(-4) + 16$
$= -12 + 16$
$= 4$

35. $12 \div (-6) + (7-2)^3 = 12 \div (-6) + 5^3$
$= 12 \div (-6) + 125$
$= -2 + 125$
$= 123$

37. $\left(-1\frac{1}{2}\right)(-4) - 3\frac{1}{4} \div \frac{1}{4} = \left(-\frac{3}{2}\right)(-4) - \frac{13}{4}\left(\frac{4}{1}\right)$
$= \left(-\frac{3}{2}\right)(-4) - \frac{13}{4}\left(\frac{4}{1}\right)$
$= 6 - 13$
$= -7$

39. $\left(\frac{3}{5} - \frac{2}{5}\right)^2 + \frac{3}{2}\left(-\frac{1}{5}\right) = \left(\frac{1}{5}\right)^2 - \frac{3}{10}$
$= \frac{1}{25} - \frac{3}{10}$
$= \frac{2}{50} - \frac{15}{50}$
$= -\frac{13}{50}$

41. $(1.2)^2 - 3.6(-1.5) = 1.44 - 3.6(-1.5)$
$= 1.44 + 5.4$
$= 6.84$

43. $\frac{-13^\circ + (-14^\circ) + (-20^\circ)}{3} = \frac{-47}{3} \approx -15.7^\circ\text{F}$

45. $\left[14^\circ + (-1)^\circ + (-13^\circ) + (-14^\circ) + (-20^\circ) + (-16^\circ)\right] \div 8$

$= \frac{50^\circ}{6}$

$= \frac{-25^\circ}{3}$

$\approx -8.3^\circ \text{F}$

47. $\left[-14^\circ + (-20^\circ) + (-16^\circ) + (-2^\circ) + 19^\circ + 33^\circ\right.$

$\left. + 39^\circ + 38^\circ + 31^\circ + 14^\circ + (-1^\circ) + (-13^\circ)\right] \div 12$

$= \frac{108^\circ}{12}$

$= 9^\circ \text{F}$

49. $-12 + 3 = -9^\circ \text{C}$

$-9 + 4(3) = -9 + 12 = 3^\circ \text{C}$

Cumulative Review

51. $3840 \text{ m} = \frac{3840}{1000} \text{ km} = 3.84 \text{ km}$

53. $A = \pi r^2$

$= 3.14(6)^2$

$= 3.14(36)$

$= 113.04 \text{ m}^2$

9.5 Exercises

1. Our number system is structured according to base 10. By making scientific notation also in base 10, the calculations are easier to perform.

3. The first part is a number greater than or equal to 1 but smaller than 10. It has at least one non-zero digit. The second part is 10 raised to some integer power.

5. $120 = 1.2 \times 10^2$

7. $1900 = 1.9 \times 10^3$

9. $26,300 = 2.63 \times 10^4$

11. $288,000 = 2.88 \times 10^5$

13. $10,000 = 1 \times 10^4$

15. $12,000,000 = 1.2 \times 10^7$

17. $0.0931 = 9.31 \times 10^{-2}$

19. $0.00279 = 2.79 \times 10^{-3}$

21. $0.82 = 8.2 \times 10^{-1}$

23. $0.000016 = 1.6 \times 10^{-5}$

25. $0.00000531 = 5.31 \times 10^{-6}$

27. $0.00000008 = 8 \times 10^{-6}$

29. $5.36 \times 10^4 = 53,600$

31. $5.334 \times 10^3 = 5334$

33. $4.6 \times 10^{12} = 4,600,000,000,000$

35. $6.2 \times 10^{-2} = 0.062$

37. $3.71 \times 10^{-1} = 0.371$

39. $9 \times 10^{11} = 900,000,000,000$

41. $3.862 \times 10^{-8} = 0.00000003862$

43. $35,689 = 3.5689 \times 10^4$

45. $5.2 \times 10^{-2} = 0.052$

47. $0.000398 = 3.98 \times 10^{-4}$

49. $1.88 \times 10^6 = 1,880,000$

51. $5,878,000,000,000 = 5.878 \times 10^{12}$ miles

53. $0.000000000000092 = 9.2 \times 10^{-14}$ liter

55. $1.25 \times 10^{13} = 12,500,000,000,000$ insects

57. $7.5 \times 10^{-5} = 0.000075$ centimeters

59. $1.4\times10^{10} = 14{,}000{,}000{,}000$ tons

61.
$$\begin{array}{r} 3.38\times10^{7} \\ +\ 5.63\times10^{7} \\ \hline 9.01\times10^{7} \end{array}$$
dollars

63.
$$\begin{array}{r} 5.87\times10^{21} \\ +\ 4.81\times10^{21} \\ \hline 10.68\times10^{21} \end{array} = 1.068\times10^{22} \text{ tons}$$

65.
$$\begin{array}{r} 4.00\times10^{8} \\ +\ 3.76\times10^{7} \\ \hline \end{array} \qquad \begin{array}{r} 40.00\times10^{7} \\ +\ 3.76\times10^{7} \\ \hline 36.24\times10^{7} \end{array}$$
$$= 3.624\times10^{8} \text{ feet}$$

67.
$$\begin{array}{r} 1.76\times10^{7} \\ -\ 1.16\times10^{7} \\ \hline 0.60\times10^{7} \end{array} = 6.0\times10^{6} \text{ square miles}$$

69. $10.5(5.88 \text{ trillion}) = 61.74 \text{ trillion}$
$= 61.74\times10^{12}$
$= 6.174\times10^{13}$ miles

Cumulative Review

71.
$$\begin{array}{r} 12.5 \\ \underline{0.21} \\ 125 \\ \underline{250} \\ 2.625 \end{array}$$

73. $\text{Cost} = 4(2\times9+2\times13)$
$= 4(18+26)$
$= 4(44)$
$= \$176$

Putting Your Skills to Work

1. $18.75-18.92=-0.17$
$17.40-18.75=-1.35$
$18.71-17.40=+1.31$
$19.52-18.71=+0.81$
$20.50-19.52=+0.98$

2. Sears: $40.62-31.98=+8.64$
Coca-Cola: $17.40-18.92=-1.52$

3. $6.10-4.65=+1.45$
$100 \text{ shares}\times\frac{\$1.45}{\text{share}} = \$145$

4. $\frac{1000}{33.70} = 29.67$, 29 shares
$\frac{\$53.25}{1 \text{ share}}\times 29 \text{ shares} = \$1544.25$
Purchase cost:
$29 \text{ share}\times\frac{\$33.70}{\text{share}} = \$977.3$
Profit: \$1544.25-\$977.30 = \$566.95

5. Krispy Kreme: $\frac{1000}{41.27} = 24.23$
Buy: $24 \text{ shares}\times\frac{\$41.27}{\text{share}} = \$990.48$
Sell: $24 \text{ shares}\times\frac{\$42.97}{\text{share}} = \$1031.28$
$1000-990.48=\$9.52$
$2000+9.52=\$2009.52$
Six Flags: $\frac{2009.52}{6.77} = 296.82$
Buy: $295 \text{ shares}\times\frac{\$6.77}{\text{share}} = \$2003.92$
Sell: $295 \text{ shares}\times\frac{\$5.42}{\text{share}} = \$1598.90$
Total Purchase: $990.48+1997.15$
$= \$2987.63$
Total Sale: $1031.28+1598.90$
$= \$2630.18$
Loss: $2987.63-2630.18$
$= \$357.45$

Chapter 9 Review Problems

1. $-20+5=-15$

2. $-18+4=-14$

3. $-3.6+(-5.2)=-8.8$

4. $10.4+(-7.8)=2.6$

5. $-\frac{1}{5}+\left(-\frac{1}{3}\right)=-\frac{3}{15}+\left(-\frac{5}{15}\right)$
$=-\frac{8}{15}$

6. $-\frac{2}{7}+\frac{5}{4}=-\frac{4}{14}+\frac{5}{14}$
$=\frac{1}{14}$

7. $20+(-14)=6$

8. $-80+60=-20$

9. $(-82)+50+35+(-18)$
$=-100+85$
$=-15$

10. $12+(-7)+(-8)+3$
$=15+(-15)$
$=0$

11. $25-36=25+(-36)=-11$

12. $12-40=12+(-40)=-28$

13. $12-(-7)=12+7=19$

14. $14-(-3)=14+3=17$

15. $-11.4-5.8=-11.4+(-5.8)=-17.2$

16. $-5.2-7.1=-5.2+(-7.1)=-12.3$

17. $-\frac{2}{5}-\left(-\frac{1}{3}\right)=-\frac{2}{5}+\frac{1}{3}$
$=-\frac{6}{15}+\frac{5}{15}$
$=-\frac{1}{15}$

18. $3\frac{1}{4}-\left(-5\frac{2}{3}\right)=3\frac{3}{12}+5\frac{8}{12}$
$=8\frac{11}{12}$

19. $5-(-2)-(-6)=5+2+6$
$=7+6$
$=13$

20. $-15-(3)+9=-15+3+9$
$=-12+9$
$=-3$

21. $9-8-6-4$
$=9+(-8)+(-6)+(-4)$
$=1+(-6)+(-4)$
$=-5+(-4)$
$=-9$

22. $-7-8-(-3)=-7+(-8)+3-15+3=-12$

23. $\left(-\frac{2}{7}\right)\left(-\frac{1}{5}\right)=\frac{2}{35}$

24. $\left(-\frac{6}{15}\right)\left(\frac{5}{12}\right)=-\frac{1}{6}$

25. $(5.2)(-1.5)=-7.8$

26. $(-3.6)(-1.2)=4.32$

27. $-60\div(-20)=3$

28. $-18\div(-3)=6$

29. $\frac{-36}{4}=-9$

30. $\dfrac{-60}{12} = -5$

31. $\dfrac{-13.2}{-2.2} = 6$

32. $\dfrac{48}{-3.2} = -15$

33. $\dfrac{-\frac{2}{5}}{\frac{4}{7}} = -\dfrac{2}{5} \div \dfrac{4}{7}$
$= -\dfrac{2}{5} \times \dfrac{7}{4}$
$= -\dfrac{7}{10}$

34. $\dfrac{-\frac{1}{3}}{-\frac{7}{9}} = -\dfrac{1}{3} \div \left(-\dfrac{7}{9}\right)$
$= -\dfrac{1}{3} \times \left(-\dfrac{7}{9}\right)$
$= \dfrac{3}{7}$

35. $3(-5)(-2) = -15(-2) = 30$

36. $(-2)(3)(-6)(-1) = -6(-6)(-1)$
$= 36(-1)$
$= -36$

37. $8 - (-30) \div 6 = 8 - (-5)$
$= 8 + 5$
$= 13$

38. $26 + (-28) \div 4 = 26 - 7 = 19$

39. $2(-6) + 3(-4) - (13)$
$= -12 + (-12) + 13$
$= -24 + 13$
$= -11$

40. $-49 \div (-7) + 3(-2) = 7 + (-6) = 1$

41. $36 \div (-12) + 50 \div (-25)$
$= -3 + (-2)$
$= -5$

42. $21 - (-30) \div 15 = 21 + 2 = 23$

43. $50 \div 25(-4) = 2(-4) = -8$

44. $-3.5 \div (-5) - 1.2$
$= 0.7 - 1.2$
$= -0.5$

45. $2.5(-2) + 3.8 = -5 + 3.8 = -1.2$

46. $\dfrac{5-9+2}{3-5} = \dfrac{-2}{-2}$
$= 1$

47. $\dfrac{4(-6)+8-2}{15-7+2} = \dfrac{-24+8-2}{15-7+2}$
$= \dfrac{-18}{10}$
$= -\dfrac{9}{5}$
$= -1\dfrac{4}{5}$

48. $\dfrac{20 \div (-5) - (-6)}{(2)(-2)(-5)} = \dfrac{-4+6}{20}$
$= \dfrac{2}{20}$
$= \dfrac{1}{10}$

49. $\dfrac{(22-4) \div (-2)}{-12-3(5)} = \dfrac{18 \div (-2)}{=12+15}$
$= \dfrac{-9}{3}$
$= -3$

50. $-3 + 4(2-6)^2 \div (-2)$
$= -3 + 4(-4)^2 \div (-2)$
$= -3 + 4(16) \div (-2)$
$= -3 + 64 \div (-2)$
$= -3 + (-32)$
$= -35$

51. $-3(12-15)^2+2^5=-3(-3)^2+2^5$
$=-3(9)+32$
$=-27+32$
$=5$

52. $-50\div(-10)+(5-3)^4$
$=-50\div(-10)+2^4$
$=-50\div(-10)+16$
$=5+16$
$=21$

53. $\frac{2}{3}-\frac{2}{5}\div\left(\frac{1}{3}\right)\left(-\frac{3}{4}\right)=\frac{2}{3}-\frac{2}{5}\left(\frac{3}{1}\right)\left(-\frac{3}{4}\right)$
$=\frac{2}{3}-\frac{6}{5}\left(-\frac{3}{4}\right)$
$=\frac{2}{3}+\frac{18}{20}$
$=\frac{40}{60}+\frac{54}{60}$
$=\frac{94}{60}$
$=\frac{47}{30}$ or $1\frac{17}{30}$

54. $\left(\frac{2}{3}\right)^2-\frac{3}{8}\left(\frac{8}{5}\right)=\frac{4}{9}-\frac{3}{8}\left(\frac{8}{5}\right)$
$=\frac{4}{9}-\frac{3}{5}$
$=\frac{20}{45}-\frac{27}{45}$
$=-\frac{7}{45}$

55. $(0.8)^2-3.2(1.6)$
$=0.64-3.2(1.6)$
$=0.64+5.12$
$=5.76$

56. $1.4(4.7-4.9)-12.8\div(-0.2)$
$=1.4(-0.2)-12.8\div(-0.2)$
$=-0.28+64$
$=63.72$

57. $4160=4.16\times10^3$

58. $3,700,000=3.7\times10^6$

59. $200,000=2\times10^5$

60. $0.007=7.0\times10^{-3}$

61. $0.0000218=2.18\times10^{-5}$

62. $0.00000763=7.63\times10^{-6}$

63. $1.89\times10^4=18,900$

64. $3.76\times10^3=3760$

65. $7.52\times10^{-2}=0.0752$

66. $6.61\times10^{-3}=0.00661$

67. $9\times10^{-7}=0.0000009$

68. $8\times10^{-8}=0.00000008$

69. $5.36\times10^{-4}=0.000536$

70. $1.98\times10^{-5}=0.0000198$

71.
$$\begin{array}{r} 5.26\times10^{11} \\ +\ 3.18\times10^{11} \\ \hline 8.44\times10^{11} \end{array}$$

72.
$$\begin{array}{r} 7.79\times10^{15} \\ +\ 1.93\times10^{15} \\ \hline 9.72\times10^{15} \end{array}$$

73.
$$\begin{array}{r} 3.42\times10^{14} \\ -\ 1.98\times10^{14} \\ \hline 1.44\times10^{14} \end{array}$$

74. $\begin{array}{r} 1.76\times10^{26} \\ -\ 1.08\times10^{26} \\ \hline 0.68\times10^{26} = 6.8\times10^{25} \end{array}$

75. $123,120,000,000,000$
$= 1.2312\times10^{14}$ drops

76. 5.983×10^{24} kilograms

77. $5280\times2,500,000,000$
$= 13,200,000,000,000$
$= 1.32\times10^{13}$ feet

78. $93,000,000 = 9.3\times10^{7}$
$9.3\times10^{7}\text{ miles}\times\dfrac{5.28\times10^{3}\text{ feet}}{1\text{ mi}}$
$= 49.104\times10^{10}$ ft
$= 4.9104\times10^{11}$ feet

79. $1.67\text{ yg} = 1.67\times10^{-24}$ grams
$0.00091\text{ yg} = 9.1\times10^{-28}$ grams

80. 0.000000000000001

81. $384.4\times10^{6} = 384,400,000$ meters

82. $-5+6+(-7) = 1(-7) = -6$ yards

Total loss of 6 yards.

83. Top of Fred's head
$-282+6 = -276$ ft
Distance to plane
$2400-(-276) = 2400+276$
$= 2676$ ft

84. $-\$18+(-\$20)+\$40 = -\$38+\$40 = \$2$

Balance is $2.

85. $\dfrac{-16^\circ+(-18^\circ)+(-5^\circ)+3^\circ+(-12^\circ)}{5}$
$= \dfrac{-48^\circ}{5}$
$= -9.6^\circ$ F

86. $2(-1)+(-2)+4(1)+2$
$= -2+(-2)+4+2$
$= -4+4+2$
$= 0+2$
$= 2$
2 points above par

How Am I Doing? Chapter 9 Test

1. $-26+15 = -11$

2. $-31+(-12) = -43$

3. $12.8+(-8.9) = 3.9$

4. $-3+(-6)+7+(-4) = -9+7+(-4)$
$= -2+(-4)$
$= -6$

5. $-5\frac{3}{4}+2\frac{1}{4} = -3\frac{2}{4} = -3\frac{1}{2}$

6. $-\frac{1}{4}+\left(-\frac{5}{8}\right) = -\frac{2}{8}+\left(-\frac{5}{8}\right)$
$= -\frac{7}{8}$

7. $-32-6 = -32+(-6) = -38$

8. $23-18 = 23+(-18) = 5$

9. $\frac{4}{5}-\left(-\frac{1}{3}\right) = \frac{4}{5}+\frac{1}{3}$
$= \frac{12}{15}+\frac{5}{15}$
$= \frac{17}{15}$
$= 1\frac{2}{15}$

10. $-50-(-7)=-50+7=-43$

11. $-2.5-(-6.5)=-2.5+6.5=4$

12. $-8.5-2.8=-8.5+(-2.8)=-11.3$

13. $\frac{1}{12}-\left(-\frac{5}{6}\right)=\frac{1}{12}+\frac{5}{6}$
$=\frac{1}{12}+\frac{10}{12}$
$=\frac{11}{12}$

14. $-15-(-15)=-15+15=0$

15. $(-20)(-6)=120$

16. $27\div\left(-\frac{3}{4}\right)=27\times\left(-\frac{4}{3}\right)$
$=-36$

17. $-40\div(-4)=10$

18. $(-9)(-1)(-2)(4)\left(\frac{1}{4}\right)$
$=9(-2)(4)\left(\frac{1}{4}\right)$
$=-18(4)\left(\frac{1}{4}\right)$
$=-72\left(\frac{1}{4}\right)$
$=-18$

19. $\frac{-39}{-13}=3$

20. $\frac{-\frac{3}{5}}{\frac{6}{7}}=-\frac{3}{5}\div\frac{6}{7}$
$=-\frac{3}{5}\times\frac{7}{6}$
$=-\frac{7}{10}$

21. $(-12)(0.5)(-3)=(-6)(-3)=18$

22. $96\div(-3)=-32$

23. $7-2(-5)=7+10=17$

24. $-2.5-1.2\div(-0.4)=-2.5-(-3)$
$=-2.5+3$
$=0.5$

25. $18\div(-3)+24\div(-12)=-6+(-2)=-8$

26. $-6(-3)-4(3-7)^2=-6(-3)-4(-4)^2$
$=-6(-3)-4(16)$
$=18-64$
$=-46$

27. $1.3+(-9.5)+2.5+(-1.5)$
$=1.3+(-9.5)+2.5+(-1.5)$
$=-8.2+2.5+(-1.5)$
$=-5.7+(-1.5)$
$=-7.2$

28. $-48\div(-6)-7(-2)^2$
$=-48\div(-6)-7(4)$
$=8+(-28)$
$=-20$

29. $\frac{3+8-5}{(-4)(6)+(-6)(3)}=\frac{3+8-5}{-24+(-18)}$
$=\frac{6}{-42}$
$=-\frac{1}{7}$

30. $\frac{5+28\div(-4)}{7-(-5)}=\frac{5+(-7)}{7+5}$
$=\frac{-2}{12}$
$=-\frac{1}{6}$

31. $80,540=8.054\times10^4$

32. $0.000007 = 7 \times 10^{-6}$

33. $9.36 \times 10^{-5} = 0.0000936$

34. $7.2 \times 10^{4} = 72,000$

35. $\dfrac{-14^\circ + (-8^\circ) + (-5^\circ) + 7^\circ + (-11^\circ)}{5}$

$= -\dfrac{-31^\circ}{5}$

$= -6.2^\circ \text{F}$

36.

$$\begin{array}{r} 2\times 5.8\times 10^{-5} \\ +\ 2\times 7.8\times 10^{-5} \\ \hline \end{array} \qquad \begin{array}{r} 11.6\times 10^{-5} \\ +\ 15.6\times 10^{-5} \\ \hline 27.2\times 10^{-5} \end{array}$$

or 2.72×10^{-4} meter

37. $58.3 - (-128.6) = 58.3 + 128.6$

$= 186.9^\circ \text{F}$

Cumulative Test for Chapters 1 - 9

1.

$$\begin{array}{r} 28,981 \\ -\ 16,598 \\ \hline 12,383 \end{array}$$

2.

$$\begin{array}{r} 127 \\ 36\overline{)34572} \\ \underline{36} \\ 97 \\ \underline{72} \\ 252 \\ \underline{252} \\ 0 \end{array}$$

3.

$$\begin{array}{r} 3\tfrac{1}{4} \\ +\ 8\tfrac{2}{3} \\ \hline \end{array} \qquad \begin{array}{r} 3\tfrac{3}{12} \\ +\ 8\tfrac{8}{12} \\ \hline 11\tfrac{11}{12} \end{array}$$

4. $1\dfrac{5}{6} \times 2\dfrac{1}{2} = \dfrac{11}{6} \times \dfrac{5}{2}$

$= \dfrac{55}{12}$

$= 4\dfrac{7}{12}$

5. 9.812456 rounds to 9.812

6.

$$\begin{array}{r} 5.820 \\ 38.964 \\ 0.571 \\ 9.305 \\ +\ 8.800 \\ \hline 63.460 \end{array}$$

or 63.46

7.

$$\begin{array}{r} 12.89 \\ \times\ 5.12 \\ \hline 2578 \\ 1289 \\ 6445 \\ \hline 65.9968 \end{array}$$

8. $\dfrac{n}{8} = \dfrac{56}{7}$

$n \times 7 = 8 \times 56$

$n \times 7 = 448$

$\dfrac{n \times 7}{7} = \dfrac{448}{7}$

$n = 64$

9. $\dfrac{n \text{ defects}}{2808 \text{ parts}} = \dfrac{7 \text{ defects}}{156 \text{ parts}}$

$156 \times n = 7 \times 2808$

$\dfrac{156 \times n}{156} = \dfrac{19,656}{156}$

$n = 126$ defects

10. $n = 0.8\% \times 38$

$n = 0.008 \times 38$

$n = 0.304$

11. 10% of what is 12?

$$\frac{10}{100} = \frac{12}{b}$$
$$10b = 100 \times 12$$
$$\frac{10b}{10} = \frac{1200}{10}$$
$$b = 120$$

12. 94 km = 94,000 m (move 3 places right)

13. $180 \text{ in.} \times \frac{1 \text{ yd}}{36 \text{ in.}} = 5 \text{ yd}$

14.
$$\begin{aligned} P &= 2(12.5) + 2(10) \\ &= 25 + 20 \\ &= 45 \text{ feet} \end{aligned}$$

15. $A = \pi r^2 = (3.14)(5)^2 = 78.5$

78.5 square meters

16. a. 300 students age 23-25

b.
500 students age 20-22
300 students age 23-25
200 students age 26-28
100 students age 29-31
1100 students over age 19

c.
400 age 17-19
200 age 26-28
100 age 29-31
700 students less than 20 or older than 25 years.

17. $\sqrt{36} + \sqrt{49} = 6 + 7 = 13$

18. $-1.2 + (-3.5) = -4.7$

19. $-\frac{1}{4} + \frac{2}{3} = -\frac{3}{12} + \frac{8}{12} = \frac{5}{12}$

20. $7 - 8 = 7 + (-18) = -11$

21. $-8 - (-3) = -8 + 3 = -5$

22.
$$\begin{aligned} 5(-3)(-1)(-2)(2) &= -15(-1)(-2)(2) \\ &= 15(-2)(2) \\ &= -30(2) \\ &= -60 \end{aligned}$$

23. $\frac{-\frac{4}{5}}{-\frac{21}{35}} = -\frac{4}{5}\left(-\frac{35}{21}\right) = \frac{4}{3} \text{ or } 1\frac{1}{3}$

24.
$$\begin{aligned} 6 - 3(-4) &= 6 - (-12) \\ &= 6 + 12 \\ &= 18 \end{aligned}$$

25. $(-20) \div (-2) + (-6) = 10 + (-6) = 4$

26.
$$\begin{aligned} \frac{(-2)(-1) + (-4)(-3)}{1 + (-4)(2)} &= \frac{2 + 12}{1 + (-8)} \\ &= \frac{14}{-7} \\ &= -2 \end{aligned}$$

27.
$$\begin{aligned} \frac{(-11)(-8) \div 22}{1 - 7(-2)} &= \frac{88 \div 22}{1 + 14} \\ &= \frac{4}{15} \end{aligned}$$

28. $86,972 = 8.6972 \times 10^4$

29. $0.0000549 = 5.49 \times 10^{-5}$

30. $3.85 \times 10^7 = 38,500,000$

31. $7 \times 10^{-5} = 0.00007$

Chapter 10

10.1 Exercises

1. A variable is a symbol, usually a letter of the alphabet, that stands for a number.

3. All the exponents for like terms must be the same. The exponent for x must be the same. The exponent for y must be the same. In this case x is raised to the second power in the frist term but y is raised to the second power in the second term.

5. $G = 5xy$: variables are G, x, y.

7. $p = \frac{4ab}{3}$: variables are p, a, b.

9. $r = 3 \times m + 5 \times n$

$r = 3m + 5n$

11. $H = 2 \times a - 3 \times b$

$H = 2a - 3b$

13. $-16x + 26x = 10x$

15. $2x - 8x + 5x = -x$

17. $-\frac{1}{2}x + \frac{3}{4}x + \frac{1}{12}x = -\frac{6}{12}x + \frac{9}{11}x + \frac{1}{12}x$

$= \frac{4}{12}x$

$= \frac{1}{3}x$

19. $x + 3x + 8 + 7 = 4x + 1$

21. $1.3x + 10 - 2.4x - 3.6$

$= 1.3x - 2.4x + 10 - 3.6$

$= -1.1x + 6.4$

23. $16x + 9y - 11 + 21x$

$= 16x + 21x + 9y - 11$

$= (16 + 21)x + 9y - 11$

$= 37x + 9y - 11$

25. $-25 + \left(2\frac{1}{2}\right)x + 15 - \left(3\frac{1}{4}\right)x$

$= \frac{5}{2}x - \frac{13}{x} - 10$

$= \left(\frac{10}{4} - \frac{13}{4}\right)x - 10$

$= -\frac{3}{4}x - 10$

27. $7a - c + 6b - 3c - 10a$

$= 7a - 10a + 6b - c - 3c$

$= (7 - 10)a + 6b + (-1 - 3)c$

$= -3a + 6b - 4c$

29. $\frac{1}{2}x + \frac{1}{7}y - \frac{3}{4}x + \frac{5}{21}y$

$= \frac{1}{2}x - \frac{3}{4}x + \frac{1}{7}y + \frac{5}{21}y$

$= \left(\frac{1}{2} - \frac{3}{4}\right)x + \left(\frac{1}{7} + \frac{5}{21}\right)y$

$= \left(\frac{2}{4} - \frac{3}{4}\right)x + \left(\frac{3}{21} + \frac{5}{21}\right)y$

$= -\frac{1}{4}x + \frac{8}{21}y$

31. $7.3x + 1.7x + 4 - 6.4x - 5.6x - 10$

$= 7.3x + 1.7x - 6.4x - 5.6x + 4 - 10$

$= -3x - 6$

33. $-7.6n + 1.2 + 11.2m - 3.7n - 8.1m$

$= -7.6n - 3.5n + 11.2m - 8.1m + 1.2$

$= -11.1n + 3.1m + 1.2$

35. a. Perimeter

$= 4x - 2 + 3x + 6 + 5x - 3$

$= 4x + 3x + 5x - 2 + 6 - 3$

$= 12x + 1$

b. It is doubled

$2(12x + 1) = 24 + 2$

Cumulative Review

37. $\frac{n}{6}=\frac{12}{15}$
$n\times15=6\times12$
$\frac{n\times15}{15}=\frac{72}{15}$
$n=4.8$

39. $6n=18$
$\frac{6n}{6}=\frac{18}{6}$
$n=3$

41. $\frac{1}{4}n=0.05$
$4\left(\frac{1}{4}n\right)=4(0.05)$
$n=0.2$

43. $\frac{n}{2500}=\frac{100}{1.25}$
$1.25n=250,000$
$\frac{1.25n}{1.25}=\frac{250,000}{1.25}$
$n=\$200,000$

10.2 Exercises

1. variable

3. $3x$ and x, $2y$ and $-3y$

5. $9(3x-2)=9(3x)+9(-2)$
$=27x-18$

7. $(-2)(x+y)=(-2)(x)+(-2)(y)$
$=-2x-2y$

9. $(-7)(1.5x-3y)=(-7)(1.5x)+(-7)(-3y)$
$=-10.5x+21y$

11. $(-3x+7y)(-10)=(-3x)(-10)+(7y)(-10)$
$=30x-70y$

13. $(6a-5b)(8)=8(6a)+8(-5b)$
$=48a-40b$

15. $(-5y-6z)(-4)=-4(-5y)+(-4)(-6z)$
$=20y+24z$

17. $4(p+9q-10)=4(p)+6(9q)+4(-10)$
$=4p+36q-40$

19. $3\left(\frac{1}{5}x+\frac{2}{3}y-\frac{1}{4}\right)$
$=3\left(\frac{1}{5}x\right)+3\left(\frac{2}{3}y\right)+3\left(-\frac{1}{4}\right)$
$=\frac{3}{5}x+2y-\frac{3}{4}$

21. $-15(-2a-3.2b+4.5)$
$=-15(-2a)+(-15)(-3.2b)+(-15)(4.5)$
$=30a+48b-67.5$

23. $(8a+12b-9c-5)(4)$
$=(8a)(4)+(12b)(4)+(-9c)(4)+(-5)(4)$
$=32a+48b-36c-20$

25. $(-2)(1.3x-8.5y-5z+12)$
$=(-2)(1.3x)+(-2)(-8.5y)$
$+(-2)(-5z)+(-2)(12)$
$=-2.6x+17y+10z-24$

27. $\frac{1}{2}\left(2x-3y+4z-\frac{1}{2}\right)$
$=\frac{1}{2}(2x)+\frac{1}{2}(-3y)+\frac{1}{2}(4z)+\frac{1}{2}\left(-\frac{1}{2}\right)$
$=x-\frac{3}{2}y+2z-\frac{1}{4}$

29. $-\frac{1}{5}(20x-30y+5z)$

$=-\frac{1}{5}(20x)+\left(-\frac{1}{5}\right)(-30y)+\left(-\frac{1}{5}\right)(5z)$

$=-4x+6y-z$

31. $p=2(l+w)=2l+2w$

33. $A=\frac{h(B+b)}{2}$

$A=\frac{bB+hb}{2}$

35. $4(5x-1)+7(x-5)=20x-4+7x-35$

$=27x-39$

37. $10(4a+5b)-8(6a+2b)$

$=40a+50b-48a-16b$

$=-8a+34b$

39. $1.5(x+2.2y)+3(2.2x+1.6y)$

$=1.5x+3.3y+6.6x+4.8y$

$=8.1x+8.1y$

41. $2(3b+c-2a)-5(a-2x+5b)$

$=6b+2c-4a-5a+10c-25b$

$=-9a-19b+12c$

43. $A=ab+ac$

$A=a(b+c)$

Hence, $a(b+c)=ab+ac$.

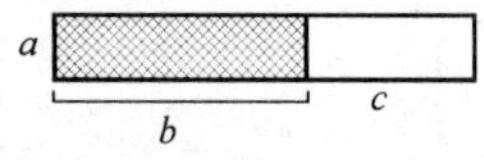

45. $A=xy+xw+xz$

$A=x(y+w+z)$

Hence, $x(y+w+z)=xy+xw+xz$.

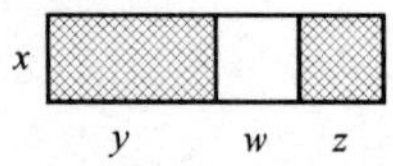

Cumulative Review

47. $P=2l+2w$

$=2(8.5)+2(5)$

$=17+10$

$=27$ in.

10.3 Exercises

1. equation

3. opposite

5. $y-12=20$

$y-12+12=20+12$

$y=32$

7. $x+6=15$

$x+6+(-6)=15+(-6)$

$x=9$

9. $x+16=-2$

$x+16+(-16)=-2+(-16)$

$x=-18$

11. $14+x=-11$

$14+(-14)+x=-11+(-14)$

$x=-25$

13. $-12+x=7$

$-12+12+x=7+12$

$x=19$

15. $5.2=x-4.6$

$5.2+4.6=x-4.6+4.6$

$9.8=x$

17. $x+3.7=-5$

$x+3.7+(-3.7)=-5+(-3.7)$

$x=-8.7$

19.
$$\begin{aligned} x - 25.2 &= -12 \\ x - 25.2 + 25.2 &= -12 + 25.2 \\ x &= 13.2 \end{aligned}$$

21.
$$\begin{aligned} \frac{4}{5} &= x + \frac{2}{5} \\ \frac{4}{5} + \left(-\frac{2}{5}\right) &= x + \frac{2}{5} + \left(-\frac{2}{5}\right) \\ \frac{2}{5} &= x \end{aligned}$$

23.
$$\begin{aligned} x - \frac{3}{5} &= \frac{2}{5} \\ x - \frac{3}{5} + \frac{3}{5} &= \frac{2}{5} + \frac{3}{5} \\ x &= \frac{5}{5} \\ &= 1 \end{aligned}$$

25.
$$\begin{aligned} x + \frac{2}{3} &= -\frac{5}{6} \\ x + \frac{2}{3} + \left(-\frac{2}{3}\right) &= -\frac{5}{6} + \left(-\frac{2}{3}\right) \\ x &= -\frac{5}{6} + \left(-\frac{4}{6}\right) \\ &= -\frac{9}{6} \\ &= -\frac{3}{2} \text{ or } -1\frac{1}{2} \end{aligned}$$

27.
$$\begin{aligned} \frac{1}{5} + y &= -\frac{2}{3} \\ \frac{1}{5} + \left(-\frac{1}{5}\right) + y &= -\frac{2}{3} + \left(-\frac{1}{5}\right) \\ y &= -\frac{10}{15} + \left(-\frac{3}{15}\right) \\ &= -\frac{13}{15} \end{aligned}$$

29.
$$\begin{aligned} 3x - 5 &= 2x - +9 \\ 3x + (-2x) - 5 &= 2x + (-2x) + 9 \\ x - 5 &= 9 \\ x - 5 + 5 &= 9 + 5 \\ x &= 14 \end{aligned}$$

31.
$$\begin{aligned} 5x + 12 &= 4x - 1 \\ 5x + (-4x) + 12 &= 4x + (-4x) - 1 \\ x + 12 &= -1 \\ x + 12 + (-12) &= -1 + (12) \\ x &= -13 \end{aligned}$$

33.
$$\begin{aligned} 7x - 9 &= 6x - 7 \\ 7x + (-6x) - 9 &= 6x + (-6x) - 7 \\ x - 9 &= -7 \\ x - 9 + 9 &= -7 + 9 \\ x &= 2 \end{aligned}$$

35.
$$\begin{aligned} 18x + 28 &= 17x + 19 \\ 18x + (-17x) + 28 &= 17 + (-17x) + 19 \\ x + 28 &= 19 \\ x + 28 + (-28) &= 19 + (-28) \\ x &= -9 \end{aligned}$$

37.
$$\begin{aligned} y + \frac{1}{2} &= 6 \\ y - \frac{1}{2} + \frac{1}{2} &= 6 + \frac{1}{2} \\ y &= 6\frac{1}{2} \end{aligned}$$

39.
$$\begin{aligned} 5 &= z + 13 \\ 5 + (-13) &= z + 13 + (-13) \\ -8 &= z \end{aligned}$$

41.
$$\begin{aligned} -5.9 + y &= -4.7 \\ -5.9 + 5.9 + y &= -4.7 + 5.9 \\ y &= 1.2 \end{aligned}$$

43. $2x-1=x+5$

$2x+(-x)-1=x+(-x)+5$

$x-1=5$

$x-1+1=5+1$

$x=6$

45. $3.6x-8=2.6x+4$

$3.6x+(-2.6x)-8=2.6x+(-2.6x)+4$

$x-8=4$

$x-8+8=4+8$

$x=12$

47. $7x+14=10x+5$

$7x+(-10x)+14=10x+(-10x)+5$

$-3x+14=5$

$-3x+14+(-14)=5+(-14)$

$-3x=-9$

$\frac{(-3x)}{-3}=\frac{(-9)}{-3}$

$x=3$

49. To solve the equation $3x=12$, divide both sides of the equation by 3 so that x stands alone on one side of the equation.

Cumulative Review

51. $5x-y+3-2x+4y=5x-2x-y+4y+3$
$=3x+3y+3$

53. $\frac{n}{100}=\frac{6.7}{142}$

$142n=100(6.7)$

$\frac{142n}{142}=\frac{670}{142}$

$n=4.7\%$

55. Mean $=\frac{18+21+24+17+21+25}{6}$

$=\frac{126}{6}$

$=21$

10.4 Exercises

1. A sample answer is: To maintain the balance, whatever you do to one side of the scale, you need to do the exact same thing to the other side of the scale.

3. $\frac{4}{3}$

5. $4x=36$

$\frac{4x}{4}=\frac{36}{4}$

$x=9$

7. $7y=-28$

$\frac{7y}{7}=\frac{-28}{7}$

$y=-4$

9. $-9y=16$

$\frac{-9y}{-9}=\frac{16}{-9}$

$y=-\frac{16}{9}$

11. $-12x=-144$

$\frac{-12x}{-12}=\frac{-144}{-12}$

$x=12$

13. $-64=-4m$

$\frac{-64}{-4}=\frac{-4m}{-4}$

$16=m$

15. $0.6x=6$

$\frac{0.6x}{0.6}=\frac{6}{0.6}$

$x=10$

17. $5.5z=9.9$

$\frac{5.5z}{5.5}=\frac{9.9}{5.5}$

$z=1.8$

19. $-0.5x = 6.75$

$$\frac{-0.5x}{0.5} = \frac{6.75}{-0.5}$$

$$x = -13.5$$

21. $\frac{5}{8}x = 3$

$$\frac{8}{5} \cdot \frac{5}{8}x = \frac{8}{5} \cdot 5$$

$$x = 8$$

23. $\frac{2}{5}y = 4$

$$\frac{2}{5} \cdot \frac{5}{2}y = 4 \cdot \frac{5}{2}$$

$$y = 10$$

25. $\frac{3}{5}n = \frac{3}{4}$

$$\frac{3}{5} \cdot \frac{5}{3}n = \frac{3}{4} \cdot \frac{5}{3}$$

$$n = \frac{5}{4}$$

$$= 1\frac{1}{4}$$

27. $\frac{3}{8}x = -\frac{3}{5}$

$$\frac{3}{8} \cdot \frac{8}{3}x = -\frac{3}{5} \cdot \frac{8}{3}$$

$$x = -\frac{8}{5}$$

$$= -1\frac{3}{5}$$

29. $\frac{1}{2}x = -2\frac{1}{4}$

$$\frac{1}{2} \cdot \frac{2}{1}x = -2\frac{1}{4} \cdot 2$$

$$x = -\frac{9}{4} \cdot 2$$

$$= -\frac{9}{2}$$

$$= -4\frac{1}{2}$$

31. $\left(-2\frac{1}{2}\right)z = -20$

$$-\frac{5}{2}z = -20$$

$$\left(-\frac{2}{5}\right)\left(-\frac{5}{2}\right)z = \left(-\frac{2}{5}\right)(-20)$$

$$z = 8$$

33. $-40 = -5x$

$$\frac{-40}{-5} = \frac{-45x}{-5}$$

$$8 = x$$

35. $\frac{2}{3}x = -6$

$$\left(\frac{3}{2}\right)\left(\frac{2}{3}x\right) = \left(\frac{3}{2}\right)(-6)$$

$$x = -9$$

37. $1.5x = 0.045$

$$\frac{1.5x}{1.5} = \frac{0.045}{1.5}$$

$$x = 0.03$$

39. $12 = -\frac{3}{5}x$

$$\left(-\frac{5}{3}\right)(12) = \left(-\frac{5}{3}\right)\left(-\frac{3}{5}x\right)$$

$$-20 = x$$

41. First, undo the multiplication:

$4.5 \div 0.5 = 9$

Now divide:

$9 \div 0.5 = 18$

18 is the correct answer.

Cumulative Review

43. $6 - 3x + 5y + 7x - 12y$
$= -3x + 7x + 5y - 12y + 6$
$= 4x - 7y + 6$

45. Increase $= \$140.50 - \$26.10$
$= \$114.40$

Percent $= \dfrac{114.40}{26.10}$
≈ 4.383
$= 438.3\%$

47. $118 + 32 = 150$ people
$30 \times 150 \times 5 = 22{,}500$ plates
$30 \times 150 \times 7 = 31{,}500$ napkins
$30 \times 150 \times 4 = 18{,}000$ towels

10.5 Exercises

1. You want to obtain the *x*-terms all by itself on one side of the equation. So you want to remove the -6 from the left-hand side of the equation. Therefore you would add the opposite of -6. This means you would add 6 to each side.

3. $3 - 4(2) \stackrel{?}{=} 5 - 3(2)$

$3 - 8 \stackrel{?}{=} 5 - 6$

$-5 \neq -1$ No

5. $8\left(\frac{1}{2}\right) - 2 \stackrel{?}{=} 10 - 16\left(\frac{1}{2}\right)$

$4 - 2 \stackrel{?}{=} 10 - 8$

$2 = 2$ Yes

7.
$$\begin{aligned} 15x - 10 &= 35 \\ 15x - 10 + 10 &= 35 + 10 \\ 15x &= 45 \\ \frac{15x}{15} &= \frac{45}{15} \\ x &= 3 \end{aligned}$$

9.
$$\begin{aligned} 6x - 9 &= -12 \\ 6x - 9 + 9 &= -12 + 9 \\ 6x &= -3 \\ \frac{6x}{6} &= \frac{-3}{6} \\ x &= -\frac{1}{2} \end{aligned}$$

11.
$$\begin{aligned} -9x &= 3x - 10 \\ -9x + (-3x) &= 3x + (-3x) - 10 \\ -12x &= -10 \\ \frac{-12x}{-12} &= \frac{-10}{-12} \\ x &= -\frac{10}{12} \\ &= \frac{5}{6} \end{aligned}$$

13.
$$\begin{aligned} -12x - 3 &= 21 \\ -12x - 3 + 3 &= 21 + 3 \\ -12x &= 24 \\ \frac{-12x}{-12} &= \frac{24}{-12} \\ x &= -2 \end{aligned}$$

15.
$$\begin{aligned} 0.26 &= 2x - 0.34 \\ 0.26 + 0.34 &= 2x - 0.34 + 0.34 \\ 0.6 &= 2x \\ \frac{0.6}{2} &= \frac{2x}{2} \\ 0.3 &= x \end{aligned}$$

17.
$$\frac{2}{3}x-5=17$$
$$\frac{2}{3}x-5+5=17+5$$
$$\frac{2}{3}x=22$$
$$\frac{2}{3}\cdot\frac{3}{2}x=22\cdot\frac{3}{2}$$
$$x=33$$

19.
$$18-2x=4x+6$$
$$18-2x+2x=4x+2x+6$$
$$18=6x+6$$
$$18+(-6)=6x+6+(-6)$$
$$12=6x$$
$$\frac{12}{6}=\frac{6x}{6}$$
$$2=x$$

21.
$$9-8x=3-2x$$
$$9-8x+8x=3-2x+8x$$
$$9=3+6x$$
$$-3+9=-3+3+6x$$
$$6=6x$$
$$\frac{6}{6}=\frac{6x}{6}$$
$$1=x$$

23.
$$2x-7=3x+9$$
$$2x+(-2x)-7=3x+(-2x)+9$$
$$-7=x+9$$
$$-7+(-9)=x+9+(-9)$$
$$-16=x$$

25.
$$1.2+0.3x=0.6x-2.1$$
$$1.2+0.3x+(-0.3x)=0.6x+(-0.3x)-2.1$$
$$1.2=0.3x-2.1$$
$$1.2+2.1=0.3x-2.1+2.1$$
$$3.3=0.3x$$
$$\frac{3.3}{0.3}=\frac{0.3x}{0.3}$$
$$11=x$$

27.
$$0.2x+0.6=-0.8-1.2x$$
$$0.2x+1.2x+0.6=-0.8-1.2x+1.2x$$
$$1.4x+0.6=-0.8$$
$$1.4x+0.6+(-0.6)=-0.8+(-0.6)$$
$$1.4x=-1.4$$
$$\frac{1.4x}{1.4}=\frac{-1.4}{1.4}$$
$$x=-1$$

29.
$$-10+6y+2=3y-26$$
$$-8+6y=3y-26$$
$$-8+6y+(-3y)=3y+(-3y)-26$$
$$-8+3y=-26$$
$$-8+8+3y=-26+8$$
$$3y=-18$$
$$\frac{3y}{3}=\frac{-18}{3}$$
$$y=-6$$

31.
$$12+4y-7=6y-9$$
$$5+4y=6y-9$$
$$5+4y+(-6y)=6y+(-6y)-9$$
$$5-2y=-9$$
$$5+(-5)-2y=-9+(-5)$$
$$-2y=-14$$
$$\frac{-2y}{-2}=\frac{-14}{-2}$$
$$y=7$$

33.
$$-30-12y+18=-24y+13+7y$$
$$-12-12y=-17y+13$$
$$-12-12y+17y=-17y+17y+13$$
$$-12+5y=13$$
$$-12+12+5y=13+12$$
$$5y=25$$
$$\frac{5y}{5}=\frac{25}{5}$$
$$y=5$$

35. $3(2x-5)-5x=1$
$$6x-15-5x=1$$
$$x-15=1$$
$$x-15+15=1+15$$
$$x=16$$

37. $5(y-2)=2(2y+3)-16$
$$5y-10=4y+6-16$$
$$5y-10=4y-10$$
$$5y+(-4y)-10=4y+(-4y)-10$$
$$y-10=-10$$
$$y-10+10=-10+10$$
$$y=0$$

39. $7x-3(x-6)=2(x-3)+8$
$$7x-3x+18=2x-6+8$$
$$4x+18=2x+2$$
$$4x+(-2x)+18=2x+(-2x)+2$$
$$2x+18=2$$
$$2x+18+(-18)=2+(-18)$$
$$2x=-16$$
$$\frac{2x}{2}=\frac{-16}{2}$$
$$x=-8$$

41. $5x+9=\frac{1}{3}(3x-6)$
$$5x+9=x-2$$
$$5x+(-x)+9=x+(-x)-2$$
$$4x+9=-2$$
$$4x+9+(-9)=-2+(-9)$$
$$4x=-11$$
$$\frac{4x}{4}=-\frac{-11}{4}$$
$$x=-\frac{11}{4}$$

43. $-2x-5(x+1)=-3(2x+5)$
$$-2x-5x-5=-6x-15$$
$$-7x-5=-6x-15$$
$$-7x+7x-5=-6x+7x-15$$
$$-5=x-15$$
$$-5+15=x-15+15$$
$$10=x$$

45. a. $7+3x=6x-8$
$$7+3x+(-6x)=6x+(-6x)-8$$
$$7-3x=-8$$
$$7+(-7)-3x=-8+(-7)$$
$$-3x=-15$$
$$\frac{-3x}{-3}=\frac{-15}{-3}$$
$$x=5$$

b. $7+3x=6x-8$
$$7+3x+(-3x)=6x+(-3x)-8$$
$$7=3x-8$$
$$7+8=3x-8+8$$
$$15=3x$$
$$\frac{15}{3}=\frac{3x}{3}$$
$$5=x$$

c. For most students collecting x-terms on the left is easier, since we usually write answers with x on the left. But either method is OK.

Cumulative Review

47. $V=\frac{4\pi r^3}{3}$
$$=\frac{4(3.14)(46\text{ cm})^3}{3}$$
$$=\frac{4(3.14)(97{,}336\text{ cu cm})}{3}$$
$$\approx 407{,}513.4\text{ cu cm}$$

49. Total area $= (18)^2 = 324 \text{ ft}^2$

Pool: $r = \frac{d}{2} = \frac{12}{2} = 6 \text{ ft}$

Area of Pool $= \pi r^2$

$= 3.14(6)^2$

$= 3.14(36)$

$= 113.04 \text{ ft}^2$

Area of deck = Total area − pool area

$= 324 - 113.04$

$= 210.96 \text{ ft}^2$

How Am I Doing? Sections 10.1 - 10.5

1. $23x - 40x = -17x$

2. $-8y + 12y - 3y = y$

3. $6a - 5b - 9a + 7b$
$= 6a - 9a - 5b + 7b$
$= -3a + 2b$

4. $5x - y + 2 - 17x - 3y + 8$
$= 5x - 17x - y - 3y + 2 + 8$
$= -12x - 4y + 10$

5. $7x - 14 + 5y + 8 - 7y + 9x$
$= 7x + 9x + 5y - 7y - 14 + 8$
$= 16x - 2y - 6$

6. $4a - 7b + 3c - 5b$
$= 4a - 7b - 5b + 3c$
$= 4a - 12b + 3c$

7. $6(7x - 3y) = 6(7x) + 6(-3y)$
$= 42x - 18y$

8. $-3(a + 5b - 1) = (-3)(a) + (-3)(5b) + (-3)(-1)$
$= -3a - 15b + 3$

9. $-2(1.5a + 3b - 6c - 5)$
$= (-2)(1.5a) + (-2)(3b) + (-2)(-6c) + (-2)(-5)$
$= -3a - 6b + 12c + 10$

10. $5(2x - y) - 3(3x + y)$
$= 5(2x) + 5(-y) + (-3)(3x) + (-3)(y)$
$= 10x - 5y - 9x - 3y$
$= 10x - 9x - 5y - 3y$
$= x - 8y$

11. $(9x + 4y)(-2) = 9x(-2) + 4y(-2)$
$= -18x - 8y$

12. $(7x - 3y)(-3) = 7x(-3) + (-3y)(-3)$
$= -21x + 9y$

13. $5 + x = 42$
$5 + (-5) + x = 42 + (-5)$
$x = 37$

14. $x + 2.5 = 6$
$x + 2.5 + (2.5) = 6 + (-2.5)$
$x = 3.5$

15. $x - \frac{5}{8} = \frac{1}{4}$
$x - \frac{5}{8} + \frac{5}{8} = \frac{1}{4} + \frac{5}{8}$
$x = \frac{2}{8} + \frac{5}{8}$
$x = \frac{7}{8}$

16. $-12 = -23 + x$
$-12 + 20 = -20 + 20 + x$
$8 = x$

17. $7x = -56$
$\frac{7x}{7} = \frac{-56}{7}$
$x = -8$

18. $5.4x = 27$
$\frac{5.4x}{5.4} = \frac{27}{5.4}$
$x = 5$

19.
$$\frac{3}{5}x = \frac{9}{10}$$
$$\frac{5}{3}\left(\frac{3}{5}x\right) = \frac{5}{3}\left(\frac{9}{10}\right)$$
$$x = \frac{3}{2} \text{ or } 1\frac{1}{2}$$

20.
$$84 = -7x$$
$$\frac{84}{-7} = \frac{-7x}{-7}$$
$$-12 = x$$

21.
$$5x - 9 = 26$$
$$5x - 9 + 9 = 26 + 9$$
$$5x = 35$$
$$\frac{5x}{5} = \frac{35}{5}$$
$$x = 7$$

22.
$$12 - 3x = 7x - 4$$
$$12 + 4 - 3x = 7x - 4 + 4$$
$$16 - 3x = 7x$$
$$16 - 3x + 3x = 7x + 3x$$
$$16 = 10x$$
$$\frac{16}{10} = \frac{10x}{10}$$
$$\frac{8}{5} = x$$

23.
$$5(x-1) = 7 - 3(x-4)$$
$$5x + 5(-1) = 7 + (-3)x + (-3)(-4)$$
$$5x - 5 = 7 - 3x + 12$$
$$5x - 5 = -3x + 19$$
$$5x - 5 + 5 = -3x + 19 + 5$$
$$5x = -3x + 24$$
$$5x + 3x = -3x + 3x + 24$$
$$8x = 24$$
$$\frac{8x}{8} = \frac{24}{8}$$
$$x = 3$$

24.
$$3x + 7 = 5(5 - x)$$
$$3x + 7 = 5(5) + 5(-x)$$
$$3x + 7 = 25 - 5x$$
$$3x + 7 + (-7) = 25 + (-7) - 5x$$
$$3x = 18 - 5x$$
$$3x + 5x = 18 - 5x + 5x$$
$$8x = 18$$
$$\frac{8x}{8} = \frac{18}{8}$$
$$x = \frac{9}{4}$$

25.
$$5x - 18 = 2(x + 3)$$
$$5x - 18 = 2x + 2(3)$$
$$5x - 18 = 2x + 6$$
$$5x - 18 + 18 = 2x + 6 + 18$$
$$5x = 2x + 24$$
$$5x - 2x = 2x - 2x + 24$$
$$3x = 24$$
$$\frac{3x}{3} = \frac{24}{3}$$
$$x = 8$$

26.
$$8x - 5(x + 2) = -3(x - 5)$$
$$8x + (-5)x + (-5)(2) = (-3x) + (-3)(-5)$$
$$8x - 5x - 10 = -3x + 15$$
$$3x - 10 = -3x + 15$$
$$3x - 10 + 10 = -3x + 15 + 10$$
$$3x = -3x + 25$$
$$3x + 3x = -3x + 3x + 25$$
$$6x = 25$$
$$\frac{5x}{6} = \frac{25}{6}$$
$$x = \frac{25}{6}$$

27.
$$\begin{aligned} 12+4y-7 &= 6y-9 \\ 4y+5 &= 6y-9 \\ 4y+5+9 &= 6y-9+9 \\ 4y+14 &= 6y \\ 4y-4y+14 &= 6y-4y \\ 14 &= 2y \\ \frac{14}{2} &= \frac{2y}{2} \\ 7 &= y \end{aligned}$$

28.
$$\begin{aligned} 0.3x+0.4 &= 0.7x-1.2 \\ 0.3x+0.4+1.2 &= 0.7x-1.2+1.2 \\ 0.3x+1.6 &= 0.7x \\ 0.3x-0.3x+1.6 &= 0.7x-0.3x \\ 1.6 &= 0.4x \\ \frac{1.6}{0.4} &= \frac{0.4x}{0.4} \\ 4 &= x \end{aligned}$$

10.6 Exercises

1. $h = 34 + r$

3. $b = n - 107$

5. $n = a + 14$

7. $l = 2w + 7$

9. $l = 3w - 2$

11. $m = 3t + 10$

13. $j + s = 26$

15. $ht = 500$

17. p = cost of the airfare to Phoenix;
$p + 135$ = cost of airfare to San Diego

19. b = number of degrees in angle B;
$b - 46$ = number of degrees in angle A

21. w = height of Mt. Whitney in meters;
$w + 4430$ = height of Mt. Everest in meters

23. a = number of books Aaron read;
$2a$ = number of books Nina read;
$a + 5$ = number of books Molly read

25. h = height;
$h + 5$ = length;
$3h$ = width

27. x = first angle;
$2x$ = second angle;
$x - 14$ = third angle

29. Speed in miles per hour of the Caravan = $1.08s$;
Speed in miles per hour of the Lexus = $1.08(s+10)$

Cumulative Review

31.
$$\begin{aligned} -6-(-7)(2) &= -6-(-14) \\ &= -6+14 \\ &= 8 \end{aligned}$$

33.
$$\begin{aligned} -2(3x+5)+12 &= 8 \\ -6x-10+12 &= 8 \\ -6x+2 &= 8 \\ -6x+2+(-2) &= 8+(-2) \\ -6x &= 6 \\ \frac{-6x}{-6} &= \frac{6}{-6} \\ x &= -1 \end{aligned}$$

35.
$$\frac{135}{349} \approx 0.387$$
$$\frac{112}{0.360} \approx 311$$
$$275 \times 0.385 \approx 106$$

10.7 Exercises

1. x = length of shorter piece

$x+5.5$ = length of longer piece

$$x+x+5.5=16$$
$$2x=10.5$$
$$\frac{2x}{2}=\frac{10.5}{2}$$
$$x=5.25$$

shorter piece = 5.25 feet

longer piece = $5.25+5.5=10.75$ feet

3. x = number of point scored by France

$x-22$ = number of point scored by Japan

$$x+x-22=80$$
$$2x-22=80$$
$$2x-22+22=80+22$$
$$2x=102$$
$$\frac{2x}{2}=\frac{102}{2}$$
$$x=51$$

51 points scored by France

$51-22=29$ points scored by Japan

5. x = number of cars in November

$x+84$ = number of cars in May

$x-43$ = number of cars in July

$$x+x+83+x-43=398$$
$$3x+40=398$$
$$3x=358$$
$$x\approx 119$$

119 cars in November

$119+84=203$ cars in May

$119-43=76$ cars in July

7. x = length of longer piece

$x-47$ = length of shorter piece

$$x+x-4.7=12$$
$$2x-4.7=12$$
$$2x=16.7$$
$$\frac{2x}{2}=\frac{16.7}{2}$$
$$x=8.35$$

longer piece = 8.35 feet

shorter piece = $8.35-4.7=3.65$ feet

9. x = width of board

$2x-4$ = length of board

$$2(x)+2(2x-4)=76$$
$$2x+4x-8=76$$
$$6x-8=76$$
$$6x=84$$
$$\frac{6x}{6}=\frac{84}{6}$$
$$x=14$$

width = 14 inches

length = $2(14)-4$

$= 28-4$

$= 24$ inches

11. x = length of the first side

$x+20$ = length of the second side

$x-4$ = length of the third side

$$x+x+20+x-4=199$$
$$3x+16=199$$
$$3x=183$$
$$x=61$$

first side = 61 mm

second side = $61+20=81$ mm

third side = $61-4=57$ mm

13. x = length of the first side

$2x$ = length of the second side

$x+12$ = length of the third side

$$x+2x+x+12=64$$
$$4x+12=44$$
$$4x=32$$
$$\frac{4x}{4}=\frac{32}{4}$$
$$x=8$$

first side = 8 cm

second side = $2(8)=16$ cm

third side = $8+12=20$ cm

15. x = number of degrees in angle A

$3x$ = number of degrees in angle B

$x+40$ = number of degrees in angle C

$$x+3x+x+40=180$$
$$5x+40=180$$
$$5x=140$$
$$\frac{5x}{5}=\frac{140}{5}$$
$$x=28$$

Angle A measures = 28°; angle B measures 84°; angle C measures = 68°

17. x = total sales

$0.05x$ = commission

$$0.05x+1200=5000$$
$$0.05x=3800$$
$$\frac{0.05x}{0.05}=\frac{3800}{0.05}$$
$$x=76{,}000$$

total sales = $76,000

19. x = yearly rent

$0.12x$ = commission

$$100+0.12x=820$$
$$0.12x=720$$
$$\frac{0.12x}{0.12}=\frac{720}{0.12}$$
$$x=6000$$

yearly rent = $6000

21. x = length of the adult section

$x+6.2$ = length of the children section

$$x+x+6.2=32$$
$$2x+6.2=32$$
$$2x=25.8$$
$$\frac{2x}{2}=\frac{25.8}{2}$$
$$x=12.9$$

Adult section = 12.9 ft

Children section = $12.9+6.2=19.1$ ft

23. x = cost of the first program

$2x-20$ = cost of the second program

$3(2x-20)+17$ = cost of the third program

$$x+2x-20+3(2x-20)+17=570.33$$
$$x+2x-20+6x-60+17=570.33$$
$$9x-63=570.33$$
$$9x=633.33$$
$$\frac{9x}{9}=\frac{633.33}{9}$$
$$x=70.37$$

first program = $70.37

second program = $2(70.37)-20$
= $120.74

third program = $3(120.74)+17$
= $379.22

25. x = value for Spain for each of 1999 and 2000.

$$\frac{5.4+6.7+x+x+7.5}{5}=6.92$$
$$\frac{19.6+2x}{5}=6.92$$
$$5\left(\frac{19.6+2x}{5}\right)=5(6.92)$$
$$19.6+2x=34.6$$
$$19.6+(-19.6)+2x=34.6+(-19.6)$$
$$2x=15$$
$$\frac{2x}{2}=\frac{15}{2}$$
$$x=7.5$$

The value is 7.5

Cumulative Review

27. What percent of 20 is 12?

$$\frac{n}{100}=\frac{12}{20}$$
$$n\times 20=100\times 12$$
$$\frac{n\times 20}{20}=\frac{1200}{20}$$
$$n=60 \text{ or } 60\%$$

29. $\frac{x}{12}=\frac{10}{15}$

$$15x=12(10)$$
$$\frac{15x}{15}=\frac{120}{15}$$
$$x=8$$

31. $C=\pi d$

$$\approx(3.14)(37.4)$$
$$=117.436$$
$$\approx 117.4 \text{ meters}$$

Putting Your Skills to Work

1. $G=2.1(19)+3.7$

$$=39.9+3.7$$
$$=43.6 \text{ seconds}$$

2.
$$27=2.1(n)+3.7$$
$$27+(-3.7)=2.1n+3.7+(-3.7)$$
$$23.3=2.1n$$
$$\frac{23.3}{2.1}=\frac{2.1n}{2.1}$$
$$11.1\approx n$$

11 vehicles per lane average

3.
$$60=2.1n+3.7$$
$$60+(-3.7)=2.1n+3.7+(-3.7)$$
$$56.3=2.1n$$
$$\frac{56.3}{2.1}=\frac{2.1n}{2.1}$$
$$26.8\approx n$$

an average of 26.8 or 27 cars per lane

4. What is 140% of 35?

$$n=1.40(35)$$
$$=49 \text{ seconds}$$

5.
$$49=2.1n+3.7$$
$$49+(-3.7)=2.1n+3.7+(-3.7)$$
$$45.3=2.1n$$
$$\frac{45.3}{2.1}=\frac{2.1n}{2.1}$$
$$21.57\approx n$$

21.57 or 22 vehicles per cycle

6. What is 155% of 35?

$$n=1.55(35)$$
$$=54.25 \text{ seconds}$$

7.
$$54.25=2.1n+3.7$$
$$54.25+(-3.7)=2.1n+3.7+(-3.7)$$
$$50.55=2.1n$$
$$\frac{50.55}{2.1}=\frac{2.1n}{2.1}$$
$$24.07\approx n$$

24 vehicles per lane per cycle

Chapter 10 Review Problems

1. $-8a+6-5a-3=-8a-5a+6-3=-13a+3$

2. $\frac{3}{4}x+\frac{2}{3}+\frac{1}{8}x+\frac{1}{4}$
$=\frac{3}{4}x+\frac{1}{8}x+\frac{2}{3}+\frac{1}{4}$
$=\frac{6}{8}x+\frac{1}{8}x+\frac{8}{12}+\frac{3}{12}$
$=\frac{7}{8}x+\frac{11}{12}$

3. $5x+2y-7x-9y$
$=5x-7x+2y-9y$
$=-2x-7y$

4. $3x-7y+8x+2y$
$=3x+8x-7y+2y$
$=11x-5y$

5. $5x-9y-12-6x-3y+18$
$=5x-6x-9y-3y-12+18$
$=-x-12y+6$

6. $7x-2y-20-5x-8y+13$
$=7x-5x-2y-8y-20+13$
$=2x-10y-7$

7. $-3(5x+y)=-3(5x)+(-3)(y)$
$=-15x-3y$

8. $-4(2x+3y)=-4(2x)+(-4)(3y)$
$=-8x-12y$

9. $2(x-3y+4)=2(x)+2(-3y)+2(4)$
$=2x-6y+8$

10. $3(2x-6y-1)=3(2x)+3(-6y)+3(-1)$
$=6x-18y-3$

11. $-15\left(\frac{1}{3}a-\frac{2}{5}b-2\right)$
$=(-15)\left(\frac{1}{3}a\right)+(-15)\left(-\frac{2}{5}b\right)+(-15)(-2)$
$=-5a+6b+30$

12. $-12\left(\frac{3}{4}a-\frac{1}{6}b-1\right)$
$=(-12)\left(\frac{3}{4}a\right)+(-12)\left(-\frac{1}{6}b\right)+(-12)(-1)$
$=-9a+2b+12$

13. $5(1.2x+3y-5.5)$
$=5(1.2x)+5(3y)+5(-5.5)$
$=6x+15y-27.5$

14. $6(1.4x-2y+3.4)$
$=6(1.4x)+6(-2y)+6(3.4)$
$=8.4x-12y+20.4$

15. $2(x+3y)-4(x-2y)$
$=2x+6y-4x+8y$
$=-2x+14y$

16. $2(5x-y)-3(x+2y)$
$=10x-2y-3x-6y$
$=7x-8y$

17. $-2(a+b)-3(2a+8)$
$=-2a-2b-6a-24$
$=-8a-2b-24$

18. $-4(a-2b)+3(5-a)$
$=-4a+8b+15-3a$
$=-7a+8b+15$

19. $x-3=9$
$x+(-3)+3=9+3$
$x=12$

20.
$$\begin{aligned} x+8.3 &= 20 \\ x+8.3+(-8.3) &= 20+(-8.3) \\ x &= 11.7 \end{aligned}$$

21.
$$\begin{aligned} -8 &= x-12 \\ -8+12 &= x-12+12 \\ 4 &= x \end{aligned}$$

22.
$$\begin{aligned} 2.4 &= x-5 \\ 2.4+5 &= x-5+5 \\ 7.4 &= x \end{aligned}$$

23.
$$\begin{aligned} 3.1+x &= -9 \\ 3.1+(-3.1)+x &= -9+(-3.1) \\ x &= -12.1 \end{aligned}$$

24.
$$\begin{aligned} 7+x &= 5.8 \\ 7+(-7)+x &= 5.8+(-7) \\ x &= -1.2 \end{aligned}$$

25.
$$\begin{aligned} x-\frac{3}{4} &= 2 \\ x+\left(-\frac{3}{4}\right)+\frac{3}{4} &= 2+\frac{3}{4} \\ x &= \frac{11}{4} = 2\frac{3}{4} \end{aligned}$$

26.
$$\begin{aligned} x+\frac{1}{2} &= 3\frac{3}{4} \\ x+\frac{1}{2}+\left(-\frac{1}{2}\right) &= 3\frac{3}{4}+\left(-\frac{1}{2}\right) \\ x &= 3\frac{3}{4}+\left(-\frac{2}{4}\right) \\ x &= 3\frac{1}{4} \end{aligned}$$

27.
$$\begin{aligned} x+\frac{3}{8} &= \frac{1}{2} \\ x+\frac{3}{8}+\left(-\frac{3}{8}\right) &= \frac{1}{2}+\left(-\frac{3}{8}\right) \\ x &= \frac{4}{8}+\left(-\frac{3}{8}\right) \\ &= \frac{1}{8} \end{aligned}$$

28.
$$\begin{aligned} x-\frac{5}{6} &= \frac{2}{3} \\ x-\frac{5}{6}+\frac{5}{6} &= \frac{2}{3}+\frac{5}{6} \\ x &= \frac{4}{6}+\frac{5}{6} \\ x &= \frac{9}{6} \\ &= \frac{3}{2} \\ &= 1\frac{1}{2} \end{aligned}$$

29.
$$\begin{aligned} 2x+20 &= 25+x \\ 2x+(-x)+20 &= 25+x+(-x) \\ x+10 &= 25 \\ x+20+(-20) &= 25+(-20) \\ x &= 5 \end{aligned}$$

30.
$$\begin{aligned} 5x-3 &= 4x-15 \\ 5x+(-4x)-3 &= 4x+(-4x)-154 \\ x-3 &= -15 \\ x-3+3 &= -15+3 \\ x &= -12 \end{aligned}$$

31.
$$\begin{aligned} 8x &= -20 \\ \frac{8x}{8} &= \frac{-20}{8} \\ x &= \frac{-20}{8} = -\frac{5}{2} \\ x &= -2\frac{1}{2} \end{aligned}$$

32.
$$\begin{aligned} -12y &= 60 \\ \frac{-12y}{-12} &= \frac{60}{-12} \\ y &= -5 \end{aligned}$$

33.
$$\begin{aligned} 1.5x &= 9 \\ \frac{1.5x}{1.5} &= \frac{9}{1.5} \\ x &= 6 \end{aligned}$$

34.
$$\begin{aligned}1.8y &= 12.6\\ \frac{1.8y}{1.8} &= \frac{12.6}{1.8}\\ y &= 7\end{aligned}$$

35.
$$\begin{aligned}-7.2x &= 36\\ \frac{-7.2x}{-7.2} &= \frac{36}{-7.2}\\ x &= -5\end{aligned}$$

36.
$$\begin{aligned}6x &= 1.5\\ \frac{6x}{6} &= \frac{1.5}{6}\\ x &= 0.25\end{aligned}$$

37.
$$\begin{aligned}\frac{3}{4}x &= 6\\ \frac{4}{3}\cdot\frac{3}{4}x &= \frac{4}{3}\cdot 6\\ x &= 8\end{aligned}$$

38.
$$\begin{aligned}\frac{2}{3}x &= \frac{5}{9}\\ \frac{3}{2}\cdot\frac{2}{3}x &= \frac{3}{2}\cdot\frac{5}{9}\\ x &= \frac{5}{6}\end{aligned}$$

39.
$$\begin{aligned}5x-3 &= 27\\ 5x-3+3 &= 27+3\\ 5x &= 30\\ \frac{5x}{5} &= \frac{30}{5}\\ x &= 6\end{aligned}$$

40.
$$\begin{aligned}8x-5 &= 19\\ 8x-5+5 &= 19+5\\ 8x &= 24\\ \frac{8x}{8} &= \frac{24}{8}\\ x &= 3\end{aligned}$$

41.
$$\begin{aligned}10-x &= -3x-6\\ 10+(-10)-x &= -3x-6+(-10)\\ -x &= -3x-16\\ -x+3x &= -3x+3x-16\\ 2x &= \frac{-16}{2}\\ x &= -8\end{aligned}$$

42.
$$\begin{aligned}7-2x &= -4x-11\\ 7+(-7)-2x &= -4x-11+(-7)\\ -2x &= -4x-18\\ -2x+4x &= -4x+4x-18\\ 2x &= -18\\ \frac{2x}{2} &= \frac{-18}{2}\\ x &= -9\end{aligned}$$

43.
$$\begin{aligned}9x-3x+18 &= 36\\ 8x+18 &= 36\\ 6x+18+(-18) &= 36+(-18)\\ 6x &= 18\\ \frac{6x}{6} &= \frac{18}{6}\\ x &= 3\end{aligned}$$

44.
$$\begin{aligned}4+2x-8 &= 12+5x+4\\ 3x-4 &= 5x+16\\ 3x+(-3x)-4 &= 5x+(-3x)+16\\ -4 &= 2x+16\\ -4+(-16) &= 2x+16+(-16)\\ -20 &= 2x\\ \frac{-20}{2} &= \frac{2x}{2}\\ -10 &= x\end{aligned}$$

45.
$$\begin{aligned}5(2x-3)&=-3+6x-8x\\5(2x)+5(-3)&=-3-2x\\10x-15&=-3-2x\\10x-15+15&=-3+15-2x\\10x&=12-2x\\10x+2x&=12-2x+2x\\12x&=12\\\frac{12x}{12}&=\frac{12}{12}\\x&=1\end{aligned}$$

46.
$$\begin{aligned}2(3x-4)&=7-2x+5x\\6x-8&=7+3x\\6x+(-3x)-8&=7+3x+(-3x)\\3x-8&=7\\3x-8+8&=7+8\\3x&=15\\\frac{3x}{3}&=\frac{15}{3}\\x&=5\end{aligned}$$

47.
$$\begin{aligned}5+2y+5(y-3)&=6(y+1)\\5+2y+5y-15&=6y+6\\7y-10&=6y+6\\7y+(-6y)-10&=6y+(-6y)+6\\y-10&=6\\y-10+10&=6+10\\y&=16\end{aligned}$$

48.
$$\begin{aligned}3+5(y+4)&=4(y-2)+3\\3+5y+20&=4y-8+3\\5y+23&=4y-5\\5y+(-4y)+23&=4y+(-4y)-5\\y+23&=-5\\y+23+(-23)&=-5+(-23)\\y&=-28\end{aligned}$$

49. $w=c+3000$

50. $a=m-18$

51. $A=3B$

52. $l=2w-3$

53. r = Roberto's salary
$r+2050$ = Michael's salary

54. x = length of the first side
$2x$ = length of the second side

55. d = number of days Dennis worked
$2d+12$ = number of days Carmen worked

56. b = number of books in old library;
$2b+450$ = number of books in new library

57. x = length of one piece
$x+6.5$ = length of other piece
$$\begin{aligned}x+x+6.5&=60\\2x+6.5&=60\\2x+6.5+(-6.5)&=60+(-6.5)\\2x&=53.5\\\frac{2x}{2}&=\frac{53.5}{2}\\x&=26.75\end{aligned}$$
one piece = 26.75 feet
other piece = $26.75+6.5=33.25$ feet

58. x = old employee's salary
$x-28$ = new employee's salary
$$\begin{aligned}x+x-28&=412\\2x-28&=412\\2x-28+28&=412+28\\2x&=440\\\frac{2x}{2}&=\frac{440}{2}\\x&=220\end{aligned}$$
old employee = \$220
new employee = $220-28=\$192$

59. x = number of customers in February
$2x$ = number of customers in March
$x+3000$ = number of customers in April

$$x+2x+x+3000=45,200$$
$$4x+3000=45,200$$
$$4x+3000+(-3000)=45,200+(-3000)$$
$$4x=42,200$$
$$\frac{4x}{4}=\frac{42,200}{4}$$
$$x=10,550$$

February = 10,550
March = 2(10,550) = 21,100
April = 10,550 + 3000 = 13,550

60. x = miles on Thursday
$x+106$ = miles on Friday
$x-39$ = miles on Saturday

$$x+x+106+x-39=856$$
$$3x+67=856$$
$$2x+67+(-67)=856+(-67)$$
$$3x=789$$
$$\frac{2x}{3}=\frac{789}{3}$$
$$x=263$$

Thursday = 263 miles
Friday = 263 + 106 = 369 miles
Saturday = 263 − 30 = 224 miles

61. w = width
$2x-3$ = length

$$72=2w+2(2w-3)$$
$$72=2w+4w-6$$
$$72=6w-6$$
$$72+6=6w+(-6)+6$$
$$78=6w$$
$$\frac{78}{6}=\frac{6w}{6}$$
$$w=13$$

width = 13 in.
length = 2(13) − 3 = 23 in.

62. x = width
$3x+2$ = length

$$2(x)+2(3x+2)=180$$
$$2x+6x+4=180$$
$$8x+4=180$$
$$8x+4+(-4)=180+(-4)$$
$$8x=176$$
$$\frac{8x}{9}=\frac{176}{8}$$
$$x=22$$

width = 22 m
length = 3(22) + 2 = 68 m

63. x = angle Z
$2x$ = angle Y
$x-12$ = angle X

$$x+2x+x-12=180$$
$$4x-12=180$$
$$4x-12+12=180+12$$
$$4x=192$$
$$\frac{4x}{4}=\frac{192}{4}$$
$$x=48$$

Angle $Z=48^\circ$
Angle $Y=2(48^\circ)=96^\circ$
Angle $X=48^\circ-12^\circ=36^\circ$

64. $x+74$ = angle A
x = angle B
$3x$ = angle C

$$x+74+x+3x=180$$
$$5x+74=180$$
$$5x+74+(-74)=180+(-74)$$
$$5x=106$$
$$\frac{5x}{5}=\frac{106}{5}$$
$$x=21.2$$

angle $A=21.2+74=95.2^\circ$
angle $B=21.2^\circ$
angle $C=3(21.2)=63.6^\circ$

65. $p = 2l + 2w$

$x = \text{length}$

$x - 67 = \text{width}$

$$346 = 2(x) + 2(x - 67)$$
$$346 = 2x + 2x + 2(-67)$$
$$346 = 4x - 134$$
$$346 + 134 = 4x - 134 + 134$$
$$480 = 4x$$
$$\frac{480}{4} = \frac{4x}{4}$$
$$120 = x$$

Length = 120 yards

Width = 120 − 67 = 53 yards

66. $x = \text{length}$

$x - 44 = \text{width}$

$$2(x) + 2(x - 44) = 288$$
$$2x + 2x - 88 = 288$$
$$4x - 88 = 288$$
$$4x - 88 + 88 = 288 + 88$$
$$4x = 376$$
$$\frac{4x}{4} = \frac{376}{4}$$
$$x = 94$$

length = 94 feet

width = 94 − 44 = 50 feet

67. $x = \text{miles for second day}$

$x + 88 = \text{miles for first day}$

$$x + x + 88 = 760$$
$$2x + 88 = 760$$
$$2x + 88 + (-88) = 760 + (-88)$$
$$2x = 672$$
$$\frac{2x}{2} = \frac{672}{2}$$
$$x = 336$$

second day = 336 miles

first day = 336 + 88 = 424 miles

68. $x = \text{first week}$

$x + 156 = \text{second week}$

$x - 142 = \text{third week}$

$$x + x + 156 + x - 142 = 800$$
$$3x + 14 = 800$$
$$3x + 14 + (-14) = 800 + (-14)$$
$$3x = 786$$
$$\frac{3x}{3} = \frac{786}{3}$$
$$x = 262$$

first week = 262

second week = 262 + 156 = 418

third week = 262 − 142 = 120

69. Commission = 600 − 200 = \$400

4% of what = 400

$$0.04 \times n = 400$$
$$\frac{0.04n}{0.04} = \frac{400}{0.04}$$
$$n = 10{,}000$$

\$10,000

70. $x = \text{cost of the furniture}$

$0.08x = \text{commission}$

$$0.08x + 1500 = 3050$$
$$0.08x + 1500 + (-1500) = 3050 + (-1500)$$
$$0.08x = 1550$$
$$\frac{0.08x}{0.08} = \frac{1550}{0.08}$$
$$x = 19{,}375$$

furniture = \$19,375

How Am I Doing? Chapter 10 Test

1. $5a - 11a = -6a$

2. $\frac{1}{3}x + \frac{5}{8}y - \frac{1}{5}x + \frac{1}{2}y$

$$= \frac{1}{3}x - \frac{1}{5}x + \frac{5}{8}y + \frac{1}{2}y$$
$$= \frac{5}{15}x - \frac{3}{15}x + \frac{5}{8}y + \frac{4}{8}y$$
$$= \frac{2}{15}x + \frac{9}{8}y$$

3. $\frac{1}{4}a-\frac{2}{3}b+\frac{3}{8}a$

$=\frac{2}{8}a+\frac{3}{8}a-\frac{2}{3}b$

$=\frac{5}{8}a-\frac{2}{3}b$

4. $6a-5b-5a-3b$

$=6a-5a-5b-3b$

$=a-8b$

5. $7x-8y+2z-9z+8y$

$=7x-8y+8y+2z-9z$

$=7x-7z$

6. $x+5y-6-5x-7y+11$

$=x-5x+5y-7y-6+11$

$=-4x-2y+5$

7. $5(12x-5y)=5(12x)+5(-5y)$

$=60x-25y$

8. $4\left(\frac{1}{2}x-\frac{5}{6}y\right)=4\left(\frac{1}{2}x\right)+4\left(\frac{5}{6}y\right)$

$=2x+\frac{10}{3}y$

9. $-1.5(3a-2b+c-8)$

$=(-1.5)(3a)+(-1.5)(-2b)$

$+(-1.5)(c)+(-1.5)(-8)$

$=-4.5a+3b-1.5c+12$

10. $2(-3a+2b)-5(a-2b)=-6a+4b-5a+10b$

$=-11a+14b$

11.

$$-5-3x=19$$
$$-5+5-3x=19+5$$
$$-3x=24$$
$$\frac{-3x}{-3}=\frac{24}{-3}$$
$$x=-8$$

12.

$$x-3.45=-9.8$$
$$x-3.45+3.45=-9.8+3.45$$
$$x=-6.35$$

13.

$$-5x+9=-4x-6$$
$$-5x+5x+9=-4x+5x-6$$
$$9=x-6$$
$$9+6=x-6+6$$
$$15=x$$

14.

$$8x-2-x=3x-9-10x$$
$$7x-2=-7x-9$$
$$7x+7x-2=-7x+7x-9$$
$$14x-2=-9$$
$$14x-2+2=-9+2$$
$$14x=-7$$
$$\frac{14x}{14}=\frac{-7}{14}$$
$$x=-\frac{1}{2}$$

15.

$$0.5x+0.6=0.2x-0.9$$
$$0.5x+0.6+(-0.6)=0.2x=-0.9(-0.6)$$
$$0.5x=0.2x-1.5$$
$$0.5x-0.2x=0.2x-0.2x-1.5$$
$$0.3x=-1.5$$
$$\frac{0.3x}{0.3}=\frac{-1.5}{0.3}$$
$$x=-5$$

16.

$$-\frac{5}{6}x=\frac{7}{12}$$
$$\left(-\frac{6}{5}\right)\left(-\frac{5}{6}\right)=\left(-\frac{6}{5}\right)\left(\frac{7}{12}\right)$$
$$x=-\frac{7}{10}$$

17. $s=f+15$

18. $n=s-15,000$

19. $\frac{1}{2}s$ = measure of the first angle;

s = measure of the second angle;

$2s$ = measure of the third angle

20. w = width;

$2w - 5$ = length

21. x = acres in Prentice farm

$3x$ = acres in Smithfield farm

$$x + 3x = 348$$
$$4x = 348$$
$$\frac{4x}{4} = \frac{348}{4}$$
$$x = 87$$

Prentice farm = 87 acres;

Smithfield farm = 261 acres

22. x = Marcia's earnings

$x - 1500$ = Sam's earnings

$$x + x - 1500 = 46,500$$
$$2x - 1500 = 46,500$$
$$2x - 1500 + 1500 = 46,500 + 1500$$
$$2x = 48,000$$
$$\frac{2x}{2} = \frac{48,000}{2}$$
$$x = 24,000$$

Marcia = \$24,000

Sam = 24,000 − 1500 = \$22,500

23. x = number of afternoon students;

$x - 24$ = number of morning students;

$x + 12$ = number of evening students

$$x + x - 24 + x + 12 = 183$$
$$3x - 12 = 183$$
$$3x - 12 + 12 = 183 + 12$$
$$3x = 195$$
$$\frac{3x}{3} = \frac{195}{3}$$
$$x = 65$$

Number of afternoon students = 65

Number of morning students = 65 − 24 = 41

Number of evening students = 65 + 12 = 77

24. x = length

$\frac{1}{2}x + 8$ = width

$$2(x) + 2\left(\frac{1}{2}x + 8\right) = 118$$
$$2x + x + 16 = 118$$
$$3x + 16 = 118$$
$$3x + 16 + (-16) = 118 + (-16)$$
$$3x = 102$$
$$\frac{3x}{3} = \frac{102}{3}$$
$$x = 34$$

length = 34 feet

width = $\frac{1}{2}(34) + 8 = 25$ feet

Cumulative Test for Chapters 1 - 10

1.
$$\begin{array}{r} 456 \\ 89 \\ 123 \\ +\ \ 79 \\ \hline 747 \end{array}$$

2.
$$\begin{array}{r} 309 \\ \times\ 35 \\ \hline 1545 \\ 927\ \\ \hline 10,815 \end{array}$$

3. 45,678,934 rounds to 45,678,900

4.
$$\frac{5}{12} \div \frac{1}{6} = \frac{5}{12} \times \frac{6}{1} = \frac{5}{2} = 2\frac{1}{2}$$

5. $3\frac{1}{4}\times 2\frac{1}{2}=\frac{13}{4}\times\frac{5}{2}$
$=\frac{65}{8}$
$=8\frac{1}{8}$

6. $9.3228\times 10^3=9322.8$

7.
$$\begin{array}{r} 4182.70 \\ -\ 3555.28 \\ \hline 627.42 \end{array}$$

8. $\frac{9}{n}=\frac{40.5}{72}$
$9\times 72=n\times 40.5$
$648=n\times 40.5$
$\frac{658}{40.5}=\frac{n\times 40.5}{40.5}$
$16=n$

9. What is 20% of 150?
$n=20\%\times 150$
$n=0.2\times 150$
$n=30$

10. 34% of what number is 1870?
$34\%\times n=1870$
$0.34\times n=1870$
$\frac{0.34\times n}{0.34}=\frac{1870}{0.34}$
$n=5500$

11. 345 mm = 0.345 m

12. 30 inches $\times\frac{\text{1 feet}}{\text{12 inches}}$ = 2.5 feet

13. $c=\pi d$
$=3.14(12)$
≈ 37.7 yd

14. $A=\frac{bh}{2}$
$=\frac{(13)(22)}{2}$
$=143$

143 square meters

15. $4-8+12-32-7$
$=4+(-8)+12+(-32)+(-7)$
$=-4+12+(32)+(-7)$
$=8+(-32)+(-7)$
$=-24+(-7)$
$=-31$

16. $(6)(-3)(-4)(2)=(-18)(-4)(2)$
$=72(2)$
$=144$

17. $\frac{1}{2}a+\frac{1}{7}b+\frac{1}{4}a-\frac{3}{14}b$
$=\frac{1}{2}a+\frac{1}{4}a+\frac{1}{7}b-\frac{3}{14}b$
$=\frac{2}{4}a+\frac{1}{4}a+\frac{2}{14}b-\frac{3}{14}b$
$=\frac{3}{4}a-\frac{1}{14}b$

18. $3x-5y-12+x-8y-12+20$
$=3x+x-5y-8y-12+20$
$=4x-13y+8$

19. $-8(2x-3y+5)$
$=(-8)(2x)+(-8)(-3y)+(8)(5)$
$=-16x+24y-40$

20. $2(3x-4y)-8(x+2y)$
$=6x-8y-8x-16y$
$=-2x-24y$

21.
$$5x-5=7x-13$$
$$5x+(-7x)-5=7x+(-7x)-13$$
$$-3x-5=-13$$
$$-2x-5+5=-13+5$$
$$-2x=-8$$
$$\frac{-2x}{-2}=\frac{-8}{-2}$$
$$x=4$$

22.
$$7-9y-12=3y+5-8y$$
$$-9y-5=-5y+5$$
$$-9y+9y-5=-5y+9y+5$$
$$-5=4y+5$$
$$-5+(-5)=4y+5+(-5)$$
$$-10=4y$$
$$\frac{-10}{4}=\frac{4y}{4}$$
$$-\frac{5}{2}=y \text{ or } y=-2\frac{1}{2}$$

23.
$$x-2+5x+3=183-x$$
$$6x+1=183-x$$
$$6x+x+1=183-x+x$$
$$7x+1=183$$
$$7x+1+(-1)=183+(-1)$$
$$7x=182$$
$$\frac{7x}{7}=\frac{182}{7}$$
$$x=26$$

24.
$$9(2x+8)=20-(x+5)$$
$$18+72=20-x-5$$
$$18x+72=-x+15$$
$$18x+x+72=-x+x+15$$
$$19x+72=15$$
$$19x+72+(-72)=15+(-72)$$
$$19x=-57$$
$$\frac{19x}{19}=\frac{-57}{19}$$
$$x=-3$$

25. p = weight of printer;
$p+322$ = weight of computer

26. f = enrollment during fall;
$f-87$ = enrollment during summer

27. x = miles driven on Thursday
$x+48$ = miles driven on Friday
$x-95$ = miles driven on Saturday
$$x+x+48+x-95=1081$$
$$3x-47=1081$$
$$3x+(-47)+47=1081+(47)$$
$$3x=1128$$
$$\frac{3x}{3}=\frac{1128}{3}$$
$$x=376$$
Thursday = 376 miles;
Friday = $376+48=424$ miles;
Saturday = $376-95=281$ miles

28. x = width
$2x+8$ = length

$$2(x)+2(2x+8)=98$$
$$2x+4x+16=98$$
$$6x+16=98$$
$$6x+16+(-16)=98+(-16)$$
$$6x=82$$
$$\frac{6x}{6}=\frac{82}{6}$$
$$x=\frac{41}{3}$$
$$=13\frac{2}{3}$$

width $=13\frac{2}{3}$ feet

length $=2x+8=2\left(\frac{41}{3}\right)+8$
$$=\frac{82}{3}+\frac{24}{3}=\frac{106}{3}$$
$$=35\frac{1}{3} \text{ feet}$$

Practice Final Examination

1. $82,367$ = Eighty-two thousand, three hundred sixty-seven

2.
$$\begin{array}{r} 13,428 \\ +16,905 \\ \hline 30,333 \end{array}$$

3. $19+23+16+45+70=173$

4.
$$\begin{array}{r} 89,071 \\ -54,968 \\ \hline 34,103 \end{array}$$

5.
$$\begin{array}{r} 78 \\ \times\ 54 \\ \hline 312 \\ 390\ \\ \hline 4212 \end{array}$$

6.
$$\begin{array}{r} 2035 \\ \times\ 107 \\ \hline 14245 \\ 20350\ \ \\ \hline 217,745 \end{array}$$

7.
$$\begin{array}{r} 158 \\ 7\overline{)1106} \\ \underline{7} \\ 40 \\ \underline{35} \\ 56 \\ \underline{56} \\ 0 \end{array}$$

8.
$$\begin{array}{r} 606 \\ 26\overline{)15,756} \\ \underline{156} \\ 156 \\ \underline{156} \\ 0 \end{array}$$

9. $3^4+20\div 4\times 2+5^2=81+10+25$
$=116$

10. $512\div 16=32$
32 miles/gallon

11. $\frac{14}{30}=\frac{14\div 2}{30\div 2}=\frac{7}{15}$

12. $3\frac{9}{11}=\frac{3\times 11+9}{11}=\frac{42}{11}$

13.
$$\begin{aligned} \frac{1}{10}+\frac{3}{4}+\frac{4}{5} &= \frac{1}{10}\times\frac{2}{2}+\frac{3}{4}\times\frac{5}{5}+\frac{4}{5}\times\frac{4}{4} \\ &= \frac{2}{20}+\frac{15}{20}+\frac{16}{20} \\ &= \frac{33}{20} \\ &= 1\frac{13}{20} \end{aligned}$$

14.
$$\begin{aligned} 2\frac{1}{3}+3\frac{3}{5} &= 2\frac{5}{15}+3\frac{9}{15} \\ &= 5\frac{14}{15} \end{aligned}$$

15.
$$\begin{array}{r} 4\frac{5}{7} \\ -\,2\frac{1}{2} \\ \hline \end{array} \qquad \begin{array}{r} 4\frac{10}{14} \\ -2\frac{7}{14} \\ \hline 2\frac{3}{14} \end{array}$$

16.
$$\begin{aligned} 1\frac{1}{4}\times 3\frac{1}{5} &= \frac{5}{4}\times\frac{16}{5} \\ &= \frac{5\times 4\times 4}{4\times 5} \\ &= 4 \end{aligned}$$

17.
$$\begin{aligned} \frac{7}{9}\div\frac{5}{18} &= \frac{7}{9}\times\frac{18}{5} \\ &= \frac{14}{5} \\ &= 2\frac{4}{5} \end{aligned}$$

18. $\dfrac{5\frac{1}{2}}{3\frac{1}{4}} = \dfrac{\frac{11}{2}}{\frac{13}{4}}$
$= \dfrac{11}{2} \times \dfrac{4}{13}$
$= \dfrac{22}{13}$
$= 1\dfrac{9}{13}$

19. $1\frac{1}{2} + 3\frac{1}{4} + 2\frac{1}{10} = 1\frac{10}{20} + 3\frac{5}{20} + 2\frac{2}{20}$
$= 6\frac{17}{20}$ miles

20. $11\frac{2}{3} \div 2\frac{1}{3} = \frac{35}{3} \div \frac{7}{3}$
$= \frac{35}{3} \times \frac{3}{7}$
$= 5$

5 packages

21. $\frac{719}{1000} = 0.719$

22. $0.86 = \frac{86}{100} = \frac{43}{50}$

23. $0.315 > 0.309$

24. $506.3782 \approx 506.38$

25. $9.6 + 3.82 + 1.05 + 7.3 = 21.77$

26.
$$\begin{array}{r} 3.610 \\ -2.853 \\ \hline 0.757 \end{array}$$

27.
$$\begin{array}{r} 1.23 \\ \times\ 0.4 \\ \hline 0.492 \end{array}$$

28.
$$\begin{array}{r} 3.69 \\ 0.24\overline{)0.8856} \\ \underline{72} \\ 165 \\ \underline{144} \\ 216 \\ \underline{216} \\ 0 \end{array}$$

29.
$$\begin{array}{r} 0.8125 \\ 16\overline{)13.0000} \\ \underline{128} \\ 20 \\ \underline{16} \\ 40 \\ \underline{32} \\ 80 \\ \underline{80} \\ 0 \end{array}$$

$\frac{13}{16} = 0.8125$

30. $0.7 + (0.2)^3 - 0.08(0.03)$
$= 0.7 + 0.008 - 0.0024$
$= 0.708 - 0.0024$
$= 0.7056$

31. $\dfrac{7000}{215} = \dfrac{7000 \div 5}{215 \div 5} = \dfrac{1400 \text{ students}}{43 \text{ faculty}}$

32. $\frac{12}{15} = \frac{17}{21}$
$12 \times 21 \stackrel{?}{=} 15 \times 17$
$252 \neq 255$ No

33. $\frac{5}{9} = \frac{n}{17}$
$5 \times 17 = 9 \times n$
$\frac{85}{9} = \frac{9 \times n}{9}$
$n \approx 9.4$

34. $\frac{3}{n} = \frac{7}{18}$

$3 \times 18 = 7 \times n$

$\frac{54}{7} = 7 \times n$

$7.7 \approx n$

35. $\frac{n}{12} = \frac{5}{4}$

$n \times 4 = 12 \times 5$

$\frac{n \times 4}{4} = \frac{60}{4}$

$n = 15$

36. $\frac{n}{7} = \frac{36}{28}$

$n \times 28 = 7 \times 36$

$\frac{n \times 28}{28} = \frac{252}{28}$

$n = 9$

37. $\frac{2000}{3} = \frac{n}{5}$

$2000 \times 5 = 3 \times n$

$\frac{10,000}{3} = \frac{3 \times n}{3}$

$n \approx \$3333.33$

38. $\frac{200}{6} = \frac{325}{n}$

$200 \times n = 6 \times 325$

$\frac{200 \times n}{200} = \frac{1950}{200}$

$n = 9.75$

9.75 inches

39. $\frac{68}{5} = \frac{4000}{n}$

$68 \times n = 5 \times 4000$

$\frac{68 \times n}{68} = \frac{20,000}{68}$

$n \approx \$294.12$ withheld

40. $\frac{18}{1.2} = \frac{24}{n}$

$18 \times n = 1.2 \times 24$

$\frac{18 \times n}{18} = \frac{28.8}{18}$

$n = 1.6$

1.6 lb of butter

41. $0.0063 = 0.63\%$

42. $\frac{17}{80} = \frac{n}{100}$

$17 \times 100 = 80 \times n$

$\frac{1700}{80} = \frac{80 \times n}{80}$

$21.25 = n$

21.25%

43. $164\% = 1.64$

44. $300 \times n = 52$

$\frac{300 \times n}{300} = \frac{52}{300}$

$n \approx 0.173$

17.3%

45. 6.3% of 4800

$6.3\% \times 4800 = n$

$0.063 \times 4800 = n$

$n = 302.4$

46. $\frac{58}{100} = \frac{145}{b}$

$58 \times b = 100 \times 145$

$\frac{58 \times b}{58} = \frac{14,500}{58}$

$b = 250$

47. 126% of 3400

$126\% \times 3400 = n$

$1.26 \times 3400 = n$

$n = 4284$

48. $11,800-0.08(11,800)=11,800-944$
$=10,856$

\$10,856

49.
$$\frac{28}{100}=\frac{1260}{b}$$
$$28\times b=100\times 1260$$
$$\frac{28\times b}{28}=\frac{126,000}{28}$$
$$b=4500$$

4500 students

50. Difference $=11.28-8.40=2.88$
$$\text{percent}=\frac{2.88}{8.40}$$
$$\approx 0.343=34.3\%$$

51. $17 \text{ qt}\times\dfrac{1 \text{ gallon}}{4 \text{ quarts}}=4.25 \text{ gallons}$

52. $3.25 \text{ tons}\times\dfrac{2000 \text{ lb}}{1 \text{ ton}}=6500 \text{ lb}$

53. $16 \text{ ft}\times\dfrac{12 \text{ in.}}{1\text{ft}}=192 \text{ in.}$

54. $5.6 \text{ km}=5600 \text{ m}$

55. $6.98 \text{ g}=0.0698 \text{ kg}$

56. $2.48 \text{ ml}=0.00248 \text{ L}$

57. $12 \text{ mi}\times\dfrac{1.61 \text{ km}}{1 \text{ mi}}=19.32 \text{ km}$

58. $0.00063182=6.3182\times 10^{-4}$

59. $126,400,000,000=1.264\times 10^{11}$

60. $0.623 \text{ cm}+0.74 \text{ cm}+0.0428 \text{ cm}$
$=0.623 \text{ cm}+0.74 \text{ cm}+0.00428 \text{ cm}$
$=1.36728 \text{ cm thick}$

61. $P=2l+2w$
$=2(6)+2(1.2)$
$=12+2.4$
$=14.4 \text{ m}$

62. $P=82+13+98+13=206 \text{ cm}$

63. $A=\dfrac{bh}{2}=\dfrac{6(1.8)}{2}=5.4 \text{ sq ft}$

64.
$$A=\frac{h(b+B)}{2}$$
$$=\frac{7.5(8+12)}{2}$$
$$=\frac{7.5(20)}{2}$$
$$=75 \text{ sq m}$$

65. $A=\pi r^2=3.14(6)^2\approx 113.04 \text{ sq m}$

66. $C=2\pi r$
$$C=2(3.14)\left(\frac{18}{2}\right)$$
$$=56.52 \text{ m}$$

67.
$$V=\frac{\pi r^2 h}{3}$$
$$=\frac{\pi(4)^2(10)}{3}$$
$$\approx 167.46 \text{ cu cm}$$

68.
$$V=\frac{12(19)(2.7)}{3}$$
$$=205.2$$

205.2 cu ft

69. Total area
= Area of square + Area of triangle
$$=s^2+\frac{bh}{2}$$
$$=(5)^2+\frac{3(5)}{2}$$
$$=32.5 \text{ sq m}$$

70. $\frac{n}{130}=\frac{30}{120}$

$n\times 120=30\times 130$

$\frac{n\times 120}{120}=\frac{3900}{120}$

$n=32.5$

71. 8 million dollars in profit

72. $9-8=1$

one million dollars

73. 50° F

74. From 1990 to 2000

75. 600 students are between 17-22 years old.

76. $10+4=14$

1400 students

77. $\text{Mean}=\frac{8+12+16+17+20+22}{6}$

≈ 15.83

$\text{Median}=\frac{16+17}{2}=16.5$

78. $\sqrt{49}+\sqrt{81}=7+9$

$=16$

79. $\sqrt{123}\approx 11.091$

80. $\text{hypotenuse}=\sqrt{9^2+12^2}$

$=\sqrt{81+44}$

$=\sqrt{225}$

$=15$ feet

81. $-8+(-2)+(-3)=-10+(-3)$

$=-13$

82. $-\frac{1}{4}+\frac{3}{8}=-\frac{1}{4}\times\frac{2}{2}+\frac{3}{8}$

$=-\frac{2}{8}+\frac{3}{8}$

$=\frac{-2+3}{8}$

$=\frac{1}{8}$

83. $9-12=9+(-12)=-3$

84. $-20-(-3)=-20+3=-17$

85. $2(-3)(4)(-1)=-6(4)(-1)$

$=-24(-1)$

$=24$

86. $-\frac{2}{3}\div\frac{1}{4}=-\frac{2}{3}\times\frac{4}{1}$

$=-\frac{8}{3}$

$=-2\frac{2}{3}$

87. $(-16)\div(-2)+(-4)=8+(-4)=4$

88. $12-3(-5)=12+15=27$

89. $7-(-3)+12\div(-6)=7+3+(-2)$

$=10+(-2)$

$=8$

90. $\frac{(-3)(-1)+(-4)(2)}{(0)(6)+(-5)(2)}=\frac{3+(-8)}{0+(-10)}$

$=\frac{-5}{-10}$

$=\frac{1}{2}$

91. $5x-3y-8x-4y$

$=5x-8x-3y-4y$

$=-3x-7y$

92. $5+2a-8b-12-6a-9b$
$=5+(-12)+2a+(-6a)$
$+(-8b)+(-9b)$
$=-7-4a-17b$

93. $-2(x-3y-5)$
$=-2(x)+(-2)(-3y)+(-2)(-5)$
$=-2x+6y+10$

94. $-2(4x+2)-3(x+3y)$
$=-8x-4-3x-9y$
$=-11x-9y-4$

95.
$$5-4x=-3$$
$$5+(-5)-4x=-3+(-5)$$
$$-4x=-8$$
$$\frac{-4x}{-4}=\frac{-8}{-4}$$
$$x=2$$

96.
$$5-2(x-3)=15$$
$$5-2x+6=15$$
$$-2x+11=15$$
$$-2x+11+(-11)=15+(-11)$$
$$-2x=4$$
$$\frac{-2x}{-2}=\frac{4}{-2}$$
$$x=-2$$

97.
$$7-2x=10+4x$$
$$7-2x+(-4x)=10+4x+(-4x)$$
$$7-6x=10$$
$$7+(-7)-6x=10+(-7)$$
$$-6x=3$$
$$\frac{-6x}{-6}=\frac{3}{-6}$$
$$x=-\frac{3}{6}$$
$$x=-\frac{1}{2}$$

98.
$$-3(x+4)=2(x-5)$$
$$-3x-12=2x-10$$
$$-3x+3x-12=2x+3x-10$$
$$-12=5x-10$$
$$-12+10=5x-10+10$$
$$-2=5x$$
$$\frac{-2}{5}=\frac{5x}{5}$$
$$-\frac{2}{5}=x$$

99. x = # of students taking math
$x+12$ = # of students taking history
$2x$ = # of students taking psychology
$$x+(x+12)+2x=452$$
$$4x+12=452$$
$$4x+12+(-12)=452+(-12)$$
$$4x=440$$
$$\frac{4x}{4}=\frac{440}{4}$$
$$x=110$$
math = 110
history = 110 + 12 = 122
psychology = 2(110) = 220

100. x = width
$2x+5$ = length
$$2(x)+2(2x+5)=106$$
$$2x+4x+10=106$$
$$6x+10=106$$
$$6x+10+(-10)=106+(-10)$$
$$6x=96$$
$$\frac{6x}{6}=\frac{96}{6}$$
$$x=16$$
width = 16 m
length = 2(16) + 5 = 37 m